AF801229

SOTHIS PRESS

Sothis Press aims to completely reimagine the public domain book with an emphasis on universal access to captivatingly designed print editions. We believe that technology will enable every single book ever published to be perpetually accessible in beautiful print form. Sothis is beauty in ubiquity.

www.sothis.press
1 Waterhouse Square, London, ECIN 2ST

Mathematical Theory of Induced Voltage in the High-Tension Magneto

Francis B. Silsbee

SOTHIS PRESS

ISBN 978-1-0365-1810-3

Mathematical Theory of Induced Voltage in the High-Tension Magneto is in the public domain.

This edition is published by Sothis Press The company is registered in England and Wales. The Company Number is 14900389.

MATHEMATICAL THEORY OF INDUCED VOLTAGE IN THE HIGH-TENSION MAGNETO

By Francis B. Silsbee

ABSTRACT

Three different circuits, representing in simplified form the essential features of the high-tension magneto, are developed and equations for the electrical performance of each are given. It is shown that by the insertion of proper electrical constants in these equations the resulting performance will be substantially the same as that of an actual magneto. Methods are suggested for the experimental determination of these constants, and the agreement between this theory and observed results is shown in certain cases.

CONTENTS

NOTATION

A, B, A_1, B_1 = constants of integration.

B_1 and B_m = magnetic flux densities.

b $=\frac{R}{2L}$

C, C_1, C_2 = electrostatic capacity.

c = skin effect constant $=\sqrt{4\pi^2 f\frac{\mu}{\rho}}$

d $=\sqrt{\frac{1}{LC}}$

E, E_1, E_2 = emf of rotation or of battery.

e = base of Napierian logarithms = 2.718 . . .

f, f_1, f_2 = frequencies.

g = length of air gap.

ih = length of iron circuit.

I_0 = current at break.

$_1, i_2$ = instantaneous currents.

i_w, i_s, i_o = instantaneous currents in winding, shunting resistance, and condenser.

j $=\sqrt{-1}$

K = fractional distance from terminal to effective center of C_d.

k = coefficient of coupling $=\sqrt{\frac{M^2}{L_1L_2}}$

k_e = coefficient of coupling of eddy current circuit.

L_{12}, L_{21} = mutual inductance of double-coil model.

L, L_1, L_2 = self-inductance.

l_0 = thickness of lamination.

M = mutual inductance.

m = dependent variable in auxiliary equations.

m_1, m_2 = roots of auxiliary equations.

p = ratio of reduction from secondary to primary.

R, R_1, R_2 = resistances.

S = shunting resistance.

s = complex ratio of flux densities in skin effect.

T, T_e = time constant $\left(\frac{L}{R}\right)$.

t = time.

v_1, v_2 = instantaneous voltage of condenser terminals.

V_s = breakdown voltage of spark gap.

V_m = maximum voltage.

X, Y = parts of complete solution of differential equations.

x $=\frac{\mu g}{h}$.

y $=\frac{\alpha_1}{\alpha_2}$.

z = impedance.

Z_1, Z_m, Z_2 = impulsive impedances $=\frac{V_m}{I_0}$.

$\alpha, \alpha_1, \alpha_2$ = real part of roots of auxiliary equations.

β, β_1, β_2 = imaginary part of roots of auxiliary equations.

δ = small difference in roots.

θ = constant of integration (angle).

μ = permeability.

ρ = resistivity.

ϕ = phase angle of impulse.

ω $=2\pi\times$ frequency.

I. INTRODUCTION

1. GENERAL

One of the problems which has received attention from the National Advisory Committee for Aeronautics is that of the high-tension magneto, and a detailed study of this device has been carried on at the Bureau of Standards under its auspices. While in the past the magneto has been developed along empirical lines, there has recently appeared a tendency to place the design of this type of apparatus upon a more definite scientific basis, and as a foundation for such a rational design at least an approximate theory of the internal phenomena which take place within it must be available. It is the purpose of the present paper to develop from several points of view such an approximate theory for one particular period in the cycle of operation and to correlate certain of the theoretical conclusions with a few experimental results.

The magneto is an exceedingly complex electrical system which serves to transform energy from mechanical through magnetic and electrical forms into heat energy in a high-voltage spark. It is probably impossible to give a complete mathematical treatment which will permit of exact computation of all the phenomena which occur in such a device. Equations can, however, be developed for various combinations of circuits which approximate more or less closely to the actual magneto, and these abstract theoretical circuits will be referred to in the following pages as "models" of the magneto. Such models are useful for two distinct purposes: (1) The change in the performance of the model under various conditions gives a qualitative indication of the corresponding behavior of the actual magneto under similar conditions; (2) a quantitative expression for any specific property of the magneto, such, for example, as the effect of eddy currents or the magnitude of the secondary capacity, may be expressed by the numerical value of the corresponding quantity in the model, and changes made upon such properties as a result of changes in the design can thus be numerically expressed and compared.

2. CYCLE OF OPERATION

The complete operation of the magneto which leads to the production of a single spark consists of a rather complex cycle of operations, which may be divided into six distinct periods, as follows:

1. The rotation of the armature, while short circuited in the magnetic field of the permanent magnets, generates in the primary winding a certain current (2 to 4 amperes) and stores in the magnetic field thus produced a certain amount of energy ($\frac{1}{2}LI^2$). At the instant at which a spark is desired the primary contacts are separated by a cam which wholly changes the electrical characteristics of the circuit.

2. The rapid decrease in primary current which results from the separation of the contacts generates a high and increasing voltage in both primary and secondary windings. This process, which may be called period 2, continues for only a few ten-thousandths of a second until the voltage across the secondary terminals has risen to a sufficiently high value to break down the gas between the electrodes of the spark plug.

3. This breakdown of the gas in the spark gap is accompanied, first, by the very rapid discharge of the condenser formed by the leads from the magneto to the spark plug. This probably takes place in an oscillatory manner, with a frequency of several million cycles per second. This discharge may be called period 3.

4. After the rapid condenser discharge is completed there is a further interval, denoted by period 4, during which the remaining magnetic energy stored in the coil is dissipated in the spark gap by the arc which continues to exist across the gap. At high speeds the cam may allow the primary contacts to close before all of the energy has been thus dissipated.

5. In this case there must be recognized a short period 5, during which the slight charge remaining in the primary condenser is dissipated through the primary contacts, the spark is extinguished, and the residual energy transferred back to the primary winding.

6. The circuit then remains substantially dead during period 6 until the armature is rotated past the position of maximum flux and period 1 of the succeeding cycle begins.

It is the purpose of the present paper to discuss only the phenomena occurring during the very short period 2 between the separation of the primary contact points and the initiation of the spark at the secondary terminals. It is the phenomenon which occurs during this period which determines whether or not a voltage sufficient to break down the spark gap is attained, and consequently whether or not a spark is produced. The function of the magneto during this period is, of course, the same as that of an induction coil, and many of the equations discussed in the present paper are directly applicable to such coils. The principal difference between the two types of device lies in the closer magnetic coupling between primary and secondary windings, the presence of greater amounts of unlaminated iron, and the rather higher frequencies of the oscillations involved in the case of the magneto.

3. HISTORICAL

The induction coil, having been widely used in electrical laboratories for many years, is the subject of a very extensive literature. Probably the first mathematical treatment of this device was due to Colley,[1] who assumed that the reaction of the secondary currents upon the primary was negligible, and on this basis worked out the primary and secondary currents and voltages for several cases.

A book by Armagnat[2] gives a very excellent discussion of the performance of induction coils and contains a bibliography of other work up to 1908. Since that time Taylor-Jones[3] has developed much more rigorously the mathematics of the induction coil, using what is referred to in this paper as the "double-coil model." He finds a very satisfactory agreement between the results of computations based upon this model and experimental results which he has obtained by the use of an electrostatic oscillograph.

Relatively little theoretical work had been done upon the high-tension magneto until the outbreak of the World War of 1914–1918 found the allied nations cut off from their principal supply of magnetos. The very excellent papers by Armagnat,[4] Young,[5] and Biffi[6] appeared in rapid succession, dealing with the general cycle of operations. The last of the three gives in considerable detail the mathematics of the double-coil model for an assumed case. Taylor-Jones[7] applied his induction coil equations to the magneto and by using in them constants obtained from the observed oscillations computed values of peak voltage agreeing within about 25 per cent with those directly observed. Campbell[8] at the National Physical Laboratory attempted to apply

[1] R. Colley, Wied. Ann., **44**, p. 109; 1891.

[2] H. Armagnat, "The theory, design, and construction of induction coils." Trans. by O. A. Kenyon. McGraw-Hill Pub. Co.; 1908.

[3] E. Taylor-Jones, Phil. Mag., **22**, p. 706, 1911; Phil. Mag., **27**, p. 565, 1914; Phil. Mag., **29**, p. 2, 1915; Phil. Mag., **30**, p. 224, 1915; Electrician, **81**, pp. 376–378, Aug. 30; pp. 396–397, Sept. 6; pp. 416–418, Sept. 13, 1918; Electrician, **82**, p. 99, 1919; Electrician, **83**, pp. 167–169, 201–202; 1919.

[4] H. Armagnat, Revue Electrique, **23**, p. 321; 1915.

[5] A. P. Young, Automobile Engineer, **7**, pp. 191, 237, 262, 298; 1917.

[6] E. Biffi, Elettrotecnica, **5**, pp. 302, 326, 386, 407; 1918.

[7] E. Taylor-Jones, Phil. Mag, **36**, p. 145; 1918.

[8] N. R. Campbell, Phil. Mag., **37**, pp. 284, 372; 1919.

Taylor-Jones's equations to the magneto but found very considerable discrepancies which he attributed to the effect of eddy currents in the iron core. In connection with this work, however, he obtained, by means of a very elaborate point-by-point method, some actual plots of the wave form of the voltage induced in the magneto by the interruption of the primary current. These most remarkable experimental results are shown in Fig. 9, and will be referred to again later in this paper.

Bairsto[9] has treated at some length the case where the magneto secondary is shunted with a resistance. In this he uses a modification of the double-coil model which is, in effect, equivalent to the closed-coil model discussed in the paper, though developed from a different point of view.

The complexities in the electric circuits of the magneto which have led to the difficulties met by previous writers in expressing the phenomena in mathematical terms may be enumerated as follows:

1. The electrostatic capacity which exists in the secondary winding, both from one to another of the various layers of the winding, from the outer layer and leads to the frame of the machine, and to some extent from the edges of the intermediate layers to the frame, very materially affect the performance of the device. The last-mentioned component of the capacity causes the current at different parts of the secondary winding to be different in magnitude.

2. The varying magnetic permeability of the iron core at different flux densities has so far resisted all attempts at mathematical treatment, and the only practical way of handling the matter is to assume a certain average permeability or average inductance which corresponds to the particular range of magnetizing force to which the iron is subjected. Since there is always an air gap of considerable reluctance in the iron circuit, the iron never approaches the condition of saturation, and consequently the changes in permeability are not as great as would appear at first sight.

3. The changing reluctance of the magnetic circuit with changing position of the armature varies the effective inductance of the winding. This effect, however, is of little importance in a study of period 2, since the total motion of the armature in the very

[9] G. E. Bairsto, Journal (Brit.) I. E. E., 58, p. 507, June, 1920.

short interval comprising the period is so slight that the inductance can be considered a constant.

4. The rotation of both the primary and secondary winding in the magnetic field generates a certain emf not present in the induction coil and distinct from that produced by the interruption of the primary circuit, and this generated voltage also changes rather rapidly with the angular position of the armature. For a discussion of period 2, however, this voltage may be considered constant and usually of negligible magnitude.

5. The magnetic leakage between the primary and secondary windings introduces a very considerable complication into the circuit and is the principal reason for the use of the double-coil model. If this magnetic leakage could be neglected, then the simpler models discussed later in the paper would be strictly applicable.

6. The eddy currents set up in the iron core by the rapidly changing flux exert a considerable influence upon the electric phenomena, and the double-coil model discussed by previous writers contains no explicit provision for taking account of such effects. These eddy current effects are the more complex inasmuch as the frequencies of the oscillations involved are so high that the distribution of the eddy currents is subject to a skin effect analogous to that occurring in conductors carrying high-frequency currents, and the eddy currents can not be completely represented by any single tertiary circuit having fixed resistance and inductance.

7. Further complications are introduced in cases where it is desired to treat the effect of resistance in parallel with the secondary terminals, such as exists when the spark plug is fouled with carbon deposits.

In the present paper it is proposed to (1) review, for the sake of completeness, the double-coil model which has been treated by various writers and to point out in some detail the various transformations of energy which occur in it during period 2; (2) to develop a new type of model involving only a single coil and condenser. This model, of course, can not fit the performance of the actual magneto as closely as does the double-coil model, but its simplicity makes it of much greater usefulness and leads to a more easy visualization of the phenomena occurring throughout the complete cycle of operation; (3) still a third model, which may be called the closed-coil model, will be treated, which indi-

61701°—22——2

cates the effects to be expected from the eddy currents in the iron and which has been experimentally found to give a good approximation to the performance of the actual magneto. A variety of methods for determining the various constants which should be inserted in these three types of models will then be discussed in some detail, and some experiments indicating the extent to which these models represent the performance of the complete machine will be described.

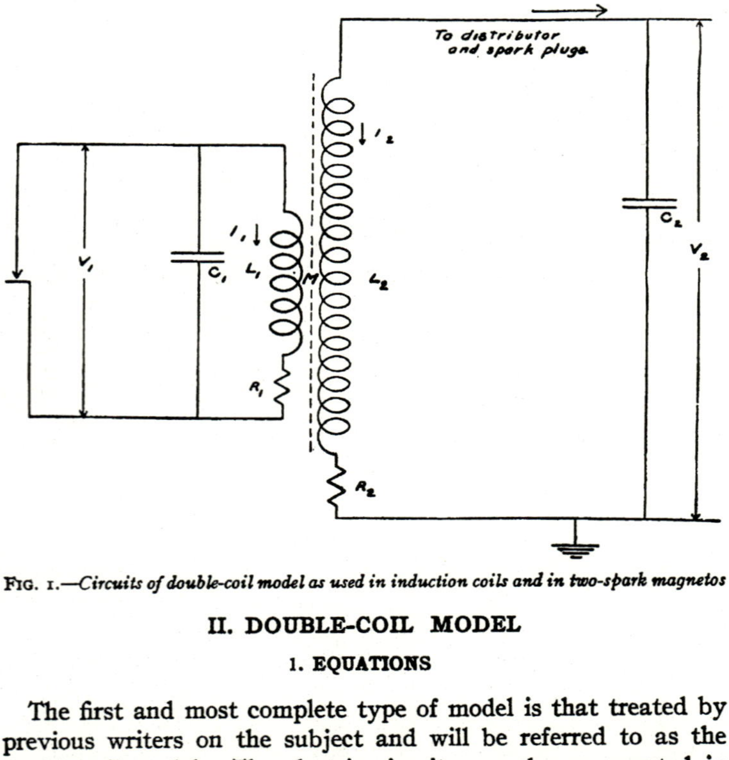

FIG. 1.—*Circuits of double-coil model as used in induction coils and in two-spark magnetos*

II. DOUBLE-COIL MODEL

1. EQUATIONS

The first and most complete type of model is that treated by previous writers on the subject and will be referred to as the double-coil model. The electric circuits may be represented in either of the two forms shown in Figs. 1 and 2. Of these, Fig. 1 corresponds to the type of circuits used in several battery ignition coils and also in so-called two-spark magnetos; i. e., magnetos intended for firing two spark plugs in the same cylinder. The second type of connection, Fig. 2, is used in certain battery systems and is widely used in magneto construction. While the equations which result from these two types of model differ slightly in detail, the general physical phenomena which take place are the same, and most of the following discussion will be confined to the type of connection shown in Fig. 1.

In either model the coils are identified with the actual physical windings of the magneto, and consequently may be considered as having the values of self and mutual inductance possessed by the actual windings. The values of these quantities will, of course, depend upon the angular position of the armature within

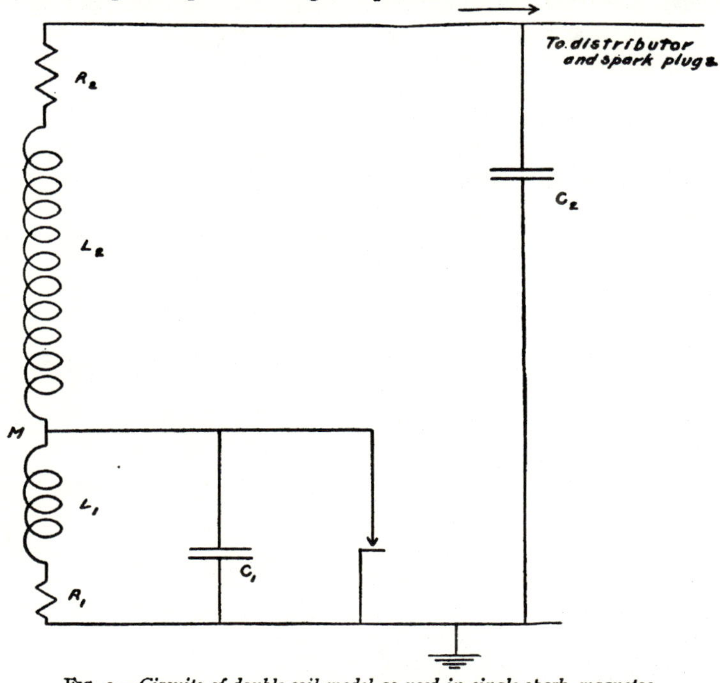

FIG. 2.—*Circuits of double-coil model as used in single-spark magnetos*

the pole pieces and upon the range of magnetizing force through which the iron is operated. It is usual to assign to the resistances R_1 and R_2 the actual resistance of the copper winding as obtained from direct-current measurements. It is possible to assign other and increased values of resistance for the purpose of representing to some degree of approximation the very greatly increased energy losses which result from the presence of eddy currents in the iron.

C_1 is identified with the capacity of the primary condenser, while C_2 represents the effect of the various secondary capacities and is best defined as that capacity which, charged to the actual terminal voltage, will contain the same amount of energy as is actually stored in electrostatic form in the dielectric between the layers and around the terminals of the actual apparatus. In the case of a magneto by far the greater part of this capacity is actually at the terminals of the winding, and consequently the current has substantially the same value throughout the length of the secondary winding. In the case of an induction coil, where the terminal capacities are relatively smaller, this condition is not entirely fulfilled, and the current is greater at the center of the winding than at the ends. Under these conditions the mutual inductance of the secondary upon the primary is not equal to the converse effect of the primary on the secondary, and we must therefore distinguish the two coefficients of mutual induction $L_{2\,1}$ and $L_{1\,2}$. This distinction can be maintained throughout the mathematics as is done by E. Taylor-Jones,[10] but the net effect is merely to reduce the coefficient of coupling defined by

$$k^2 = \frac{L_{2\,1}L_{1\,2}}{L_2L_1}$$

If we let i_1 and i_2 be the instantaneous currents in the primary and secondary circuits of Fig. 1, and v_1' and v_2' the corresponding instantaneous voltages at the terminals of the condensers, and also E_1 and E_2 be the emf's induced in the respective windings by the rotation of the armature, we have, applying Kirchhoff's laws to the two circuits:

$$L_1\frac{di_1}{dt} + M\frac{di_2}{dt} + R_1i_1 + v_1' = E_1 \qquad (1)$$

$$L_2\frac{di_2}{dt} + M\frac{di_1}{dt} + R_2i_2 + v_2' = E_2 \qquad (2)$$

We also have the relations:

$$i_1 = C_1\frac{dv_1'}{dt} \qquad (3)$$

$$i_2 = C_2\frac{dv_2'}{dt} \qquad (4)$$

[10] E. Taylor-Jones, Phil. Mag., **36**, p. 145: 1918.

Combining these gives:

$$L_1C_1\frac{d^2v_1'}{dt^2}+MC_2\frac{d^2v_2'}{dt^2}+C_1R_1\frac{dv_1'}{dt}+v_1'=E_1 \tag{5}$$

$$L_2C_2\frac{d^2v_2'}{dt^2}+MC_1\frac{d^2v_1'}{dt^2}+C_2R_2\frac{dv_2'}{dt}+v_2'=E_2 \tag{6}$$

If we further let $v_1=v_1'-E_1$, and $v_2=v_2'-E_2$ and note that the rates of change of E_1 and E_2 are negligibly small compared with the exceedingly rapid rates of change of v_1 and v_2, we obtain: (7)

$$L_1C_1\frac{d^2v_1}{dt^2}+MC_2\frac{d^2v_2}{dt^2}+C_1R_1\frac{dv_1}{dt}+v_1=0 \tag{8}$$

$$L_2C_2\frac{d^2v_2}{dt^2}+MC_1\frac{d^2v_1}{dt^2}+C_2R_2\frac{dv_2}{dt}+v_2=0 \tag{9}$$

These two simultaneous differential equations may be combined to eliminate one of the variables and give the following final equation:

$$(L_1L_2-M^2)C_1C_2\frac{d^4v_2}{dt^4}+(L_1R_2+L_2R_1)C_1C_2\frac{d^3v_2}{dt^3}$$
$$+(R_1R_2C_1C_2+L_1C_1+L_2C_2)\frac{d^2v_2}{dt^2}+(R_1C_1+R_2C_2)\frac{dv_2}{dt}+v_2=0 \tag{10}$$

The corresponding equation for the type of circuit shown in Fig. 2 is

$$(L_1L_2-M^2)C_1C_2\frac{d^4v_2}{dt^4}+(L_1R_2+L_2R_1)C_1C_2\frac{d^3v_2}{dt^3}$$
$$+\{R_1R_2C_1C_2+L_1C_1+(L_2+2M+L_1)C_2\}\frac{d^2v_2}{dt^2} \tag{11}$$
$$+\{R_1C_1+(R_1+R_2)C_2\}\frac{dv_2}{dt}+v_2=0$$

These equations will be immediately recognized as the usual ones for coupled circuits so widely used in radiotelegraphy, and their application in this case differs from the more usual treatment only in the different initial condition which must be satisfied. These equations have been treated by numerous writers, among whom may be mentioned Drude,[11] Cohen,[12] and Pernot.[13] The author last named has contributed a very detailed discussion of the equations with much practical advice for their solution in numerical cases.

---

[11] P. Drude, Ann. Physik., 13, p. 512:: 1904.
[12] L. Cohen, Bull. Bureau of Standards, 5, p. 511:: 1909.
[13] F. E. Pernot, University of California Publications in Engineering, 1, p. 161:: 1916.

The general solution[14] for equation (10) is

$$v_2 = A_1e^{m_1t} + A_2e^{m_2t} + A_3e^{m_3t} + A_4e^{m_4t} \tag{12}$$

where m_1, m_2, m_3, m_4 are the roots of the biquadratic equation:

$$m^4 + \frac{L_1R_2 + L_2R_1}{L_1L_2 - M^2}m^3 + \frac{R_1R_2 + \frac{L_1}{C_2} + \frac{L_2}{C_1}}{L_1L_2 - M^2}m^2 + \frac{\frac{R_1}{C_2} + \frac{R_2}{C_1}}{L_1L_2 - M^2}m + \frac{1}{(L_1L_2 - M^2)C_1C_2} = 0 \tag{13}$$

Since all the coefficients in this equation are essentially positive, it may be shown from the theory of equations that (except for limiting cases) only three types of solution arise: (*a*) All the roots are real and negative; (*b*) two roots are real and negative, while the other two are complex quantities; (*c*) all the roots are complex and form two conjugate pairs. In cases (*b*) and (*c*) the real part of the complex roots is zero or negative. Most cases of actual magneto circuits come under case (*c*), and for this case the results are more conveniently expressed in the form

$$v_2 = e^{\alpha_1 t}(A_1 \sin \beta_1 t + B_1 \cos \beta_1 t) + e^{\alpha_2 t}(A_2 \sin \beta_2 t + B_2 \cos \beta_2 t) \tag{14}$$

where A_1, A_2, B_1, and B_2 are constants of integration, and $\alpha_1 \pm j\beta_1$ and $\alpha_2 \pm j\beta_2$ are the roots of equation (13). The equations for v_1, i_1, and i_2 are in the same form but with different integration constants.

Equation (14) shows that v_2 is the sum of two oscillations of frequency,

$$f_1 = \frac{\beta_1}{2\pi} \text{ and } f_2 = \frac{\beta_2}{2\pi}$$

The resulting variation of voltage and current with time is shown in Figs. 3, 5, and 7, and Curve I of Fig. 31 for typical cases. If the damping is negligible, the two frequencies are given by—

$$f^2 = \left(\frac{1}{2\pi}\right)^2 \frac{L_1C_1 + L_2C_2 \pm \sqrt{(L_1C_1 + L_2C_2)^2 - 4(L_1L_2 - M^2)C_1C_2}}{2(L_1L_2 - M^2)\, C_1C_2} \tag{15}$$

It may be seen on expanding the numerator by the binominal theorem that as the closeness of coupling is increased, f_2, corresponding to the negative root, remains approximately constant, while the other frequency, f_1, becomes larger and larger, varying roughly in proportion to $\frac{1}{L_1L_2 - M^2}$, and the corresponding term

[14] See Murray, Differential Equations,' p. 64, 2d ed.

in equation (14) decreases in amplitude roughly in proportion to the increase in frequency. Figs. 4, 6, and 8 show for three typical sets of values the corresponding pulsations in the energy stored in various forms in the several parts of the circuit.

2. NUMERICAL EXAMPLES

Case I corresponds to an induction coil or a very loosely coupled magneto in which the quantity k^2 has the value 0.60. The resultant curves of current and voltage show very markedly the

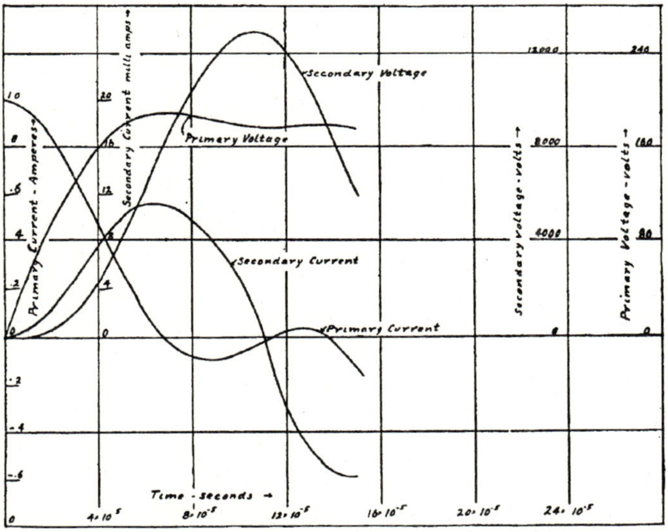

FIG. 3.—*Voltage and current wave forms during period 2 computed for double-coil model (Case I) having* $R_1=0.5\ \Omega$, $L_1=0.015\ h$, $C_1=0.2\ \mu f$, $R_2=2500\ \Omega$, $L_2=36\ h$, $C_2=50\times10^{-6}\ \mu f$, $k^2=0.60$

two oscillations which are present in such cases. It may be noted that the time $t=11\times10^{-5}$ seconds, at which the secondary voltage V_2 is a maximum, corresponds to a time when the primary voltage V_1 is a minimum, and Fig. 4 shows that the magnetic energy—i. e., the difference between total energy and total static energy—is very small at this instant. Consequently, conditions are very favorable for the production of a high secondary voltage, and a relatively large part of the total energy supplied is available for charging the secondary condenser and producing across its terminals a high voltage.

In contrast to this we may consider Case II, which corresponds to the same circuits except that the mutual inductance between them has been increased so as to make the quantity k^2 equal 0.80. Even such an arrangement is still decidedly more loosely coupled than are the circuits in the average magneto. As shown in Fig. 5, the change in coupling has somewhat shifted the relative phase of the two oscillations, so that the maximum of the secondary volt-

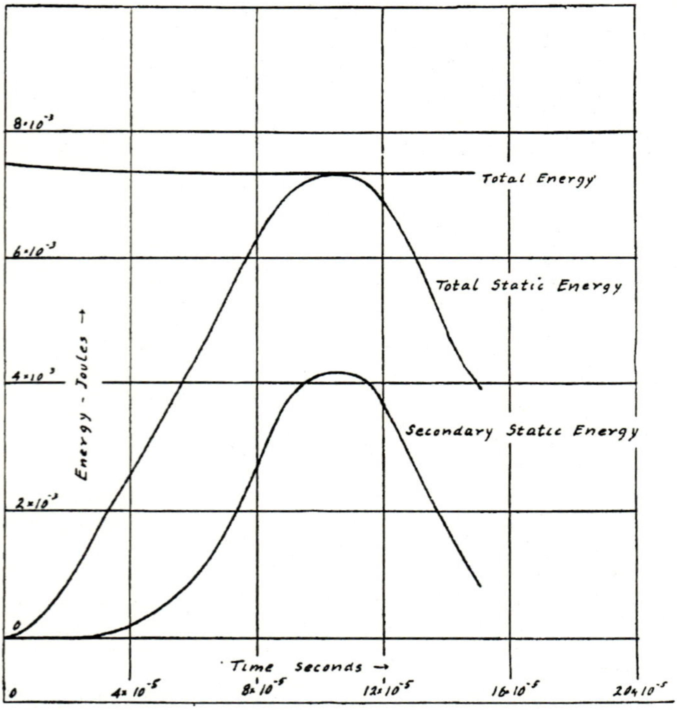

FIG. 4.—*Changes of energy with time. Double-coil model, Case I*

age no longer corresponds to a minimum of the primary voltage, and therefore at the time when the secondary voltage is a maximum ($t=8\times10^{-5}$ seconds) there is a very considerable quantity of energy stored in the magnetic circuit and later in the cycle ($t=11\times10^{-5}$ seconds) when this energy has been transferred to the primary condenser, some of the secondary energy has also returned to the primary, and consequently there is less available for charging the secondary condenser. It is this type of variation in the relative time at which the component oscillations have their maximum value which gives rise to the curious changes in maximum voltage shown by the curves obtained by Taylor-Jones. The curves of Figs. 3 and 5 also bring out the fact that the higher frequency component has a larger relative amplitude in the current waves than in the voltage waves. In both of these models the resistance has been taken to correspond with the actual resistance of the copper windings, and consequently the loss

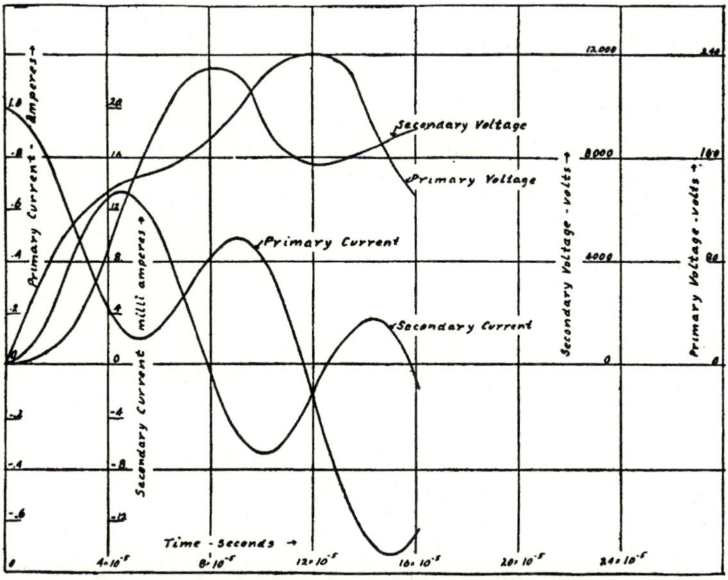

Fig. 5.—*Voltage and current wave forms during period 2 computed for double-coil model (Case II). This case is the same as Case I except that $k^2=0.80$*

of energy as shown by the total energy curves, Figs. 4 and 6, is practically negligible, and the entire energy oscillates between the three storage regions; i. e., the coil, the primary condenser, and the secondary condenser.

Case III corresponds to the same system of circuits arranged with still closer coupling, so that $k^2=0.92$. This value is typical of many magnetos. Furthermore, the resistances of both windings have been arbitrarily increased to 50 times the value of the copper resistance in order to represent as well as possible the effect of the iron losses. It will be seen that the principal effect of the increased closeness of coupling has been to greatly increase the frequency of the shorter period component and to materially reduce its amplitude, so that in the voltage wave it is hardly noticeable. The increase in resistance causes a rapid damping of the vibrations, and this damping is much greater for the higher frequency component, so that by the time $t = 10 \times 10^{-6}$ seconds, at which the secondary voltage has approached its maximum, the higher frequency component is practically negligible. The curves of total energy, Fig. 8, show that even with the increased damping

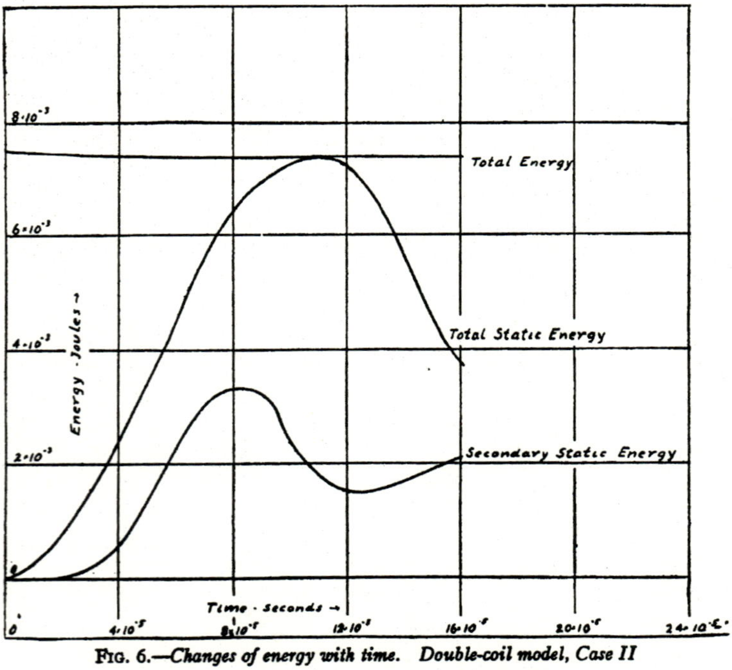

FIG. 6.—*Changes of energy with time. Double-coil model, Case II*

which results from the increased resistance the energy loss is still rather small, though it has become appreciable at the end of a complete cycle of the main wave. It is also to be noted that the primary voltage has substantially the same wave shape and is substantially in phase with the secondary voltage, and that the distribution of energy between the two at their maximum depends upon the relative capacities and turns in the windings. A comparison between Fig. 7 and Fig. 9 shows that the model for Case III, with the assumed increased resistance, roughly corresponds

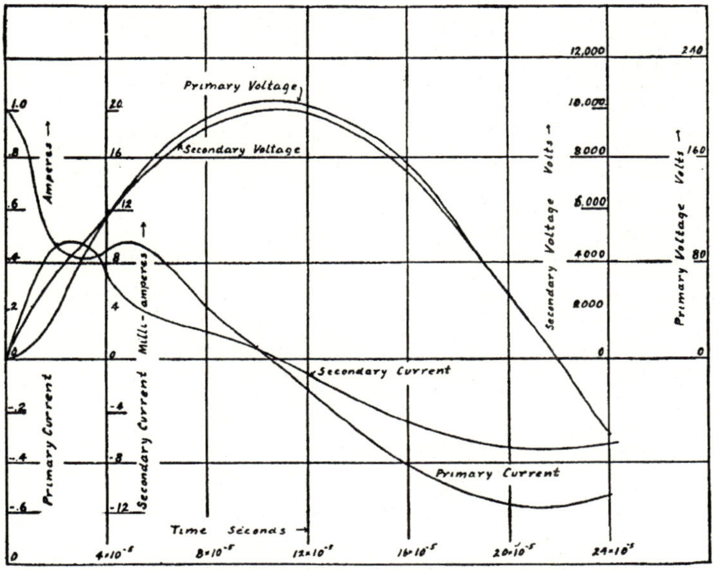

FIG. 7.—*Voltage and current wave forms during period 2 computed for double-coil model (Case III) having* $R_1=25\ \Omega$, $L_1=0.015\ h$, $C_1=0.2\ \mu f$, $R_2=125\ 000\ \Omega$, $L_2=36$, $C_2=50\times10^{-6}\ \mu f$, $k^2=0.92$

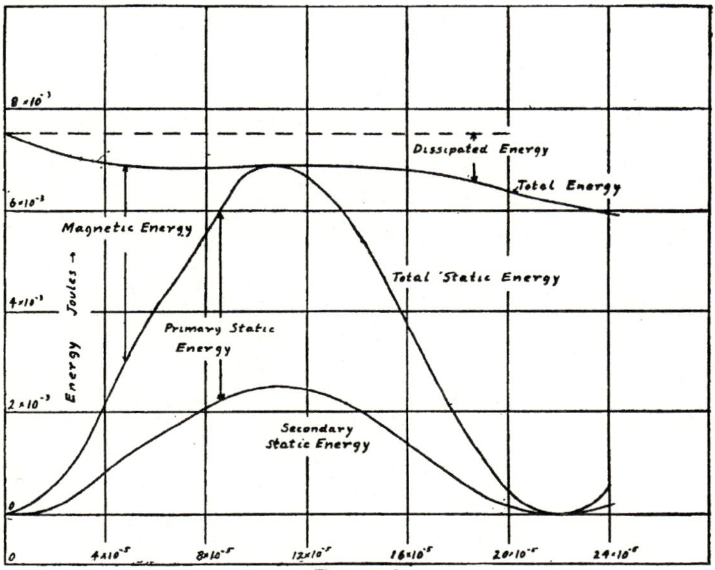

Time—second.

FIG. 8.—*Changes of energy with time. Double-coil model, Case III*

to the actual wave form as observed by Campbell, although the damping of the higher frequency ripple is slightly less and that of the lower frequency wave is much greater in his case.

The advantages of the double-coil model which has just been discussed are principally that it shows the effect of loose coupling between the primary and secondary. It indicates the two vibration frequencies which are present and permits the computation of their values. It is also possible with this model to take account of the fact that the current is not uniform throughout the length

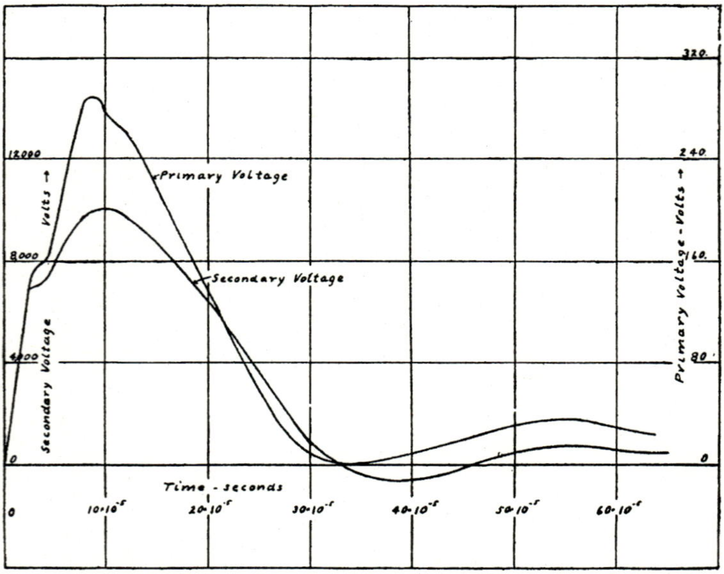

FIG. 9.—*Reproduction of wave form of shuttle type magneto as observed by Campbell. (From British Advisory Committee for Aeronautics Report, I. C. E. 241)*

of the secondary winding. It is also possible with this double-coil model to handle cases in which a resistance R_4 exists in parallel with the secondary terminals, as would be the case with a fouled spark plug. The equations for this circuit, with the connections as in Fig. 2, are

$$L_1C_1\frac{d^2v_1'}{dt^2}+(L_1+M)\,C_2\frac{d^2v_2'}{dt^2}+R_1C_1\frac{dv_1'}{dt}+$$
$$\left(\frac{L_1+M}{R_4}+R_1C_2\right)\frac{dv_2'}{dt}+\frac{R_1}{R_4}\,v_2'+v_1'=E_1 \qquad (16)$$

$$\text{and } (L_2+M)\,C_2\,\frac{d^2v_2'}{dt^2}+MC_1\,\frac{d^2v_1'}{dt^2}+\left(\frac{L_2+M}{R_4}+R_2C_2\right)\frac{dv_2'}{dt} +\left(\frac{R_2}{R_4}+1\right)v_2'-v_1'=E_2 \qquad (17)$$

The great disadvantage of this double-coil model is the excessive labor of the mathematical computations which are required in any

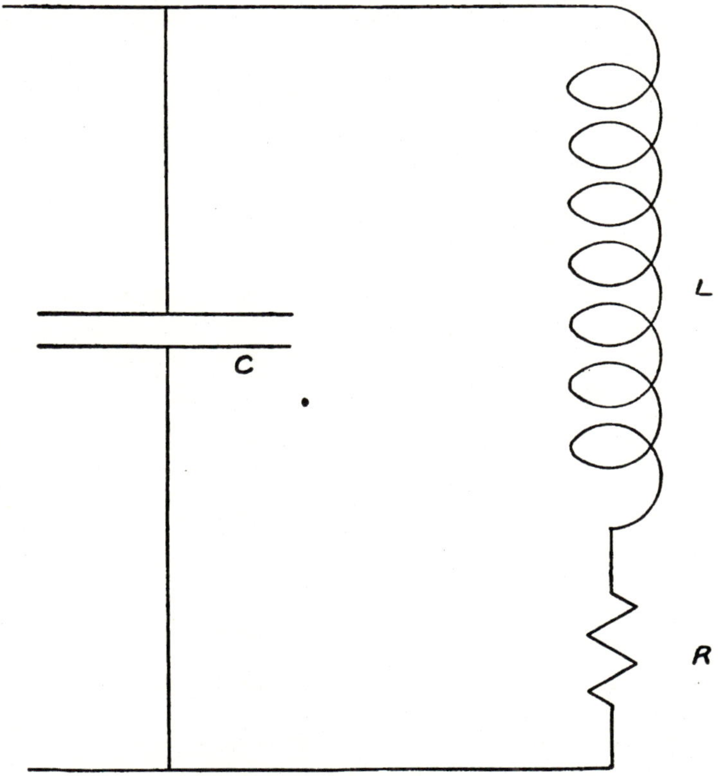

FIG. 10.—*Single-coil model circuit*

use of it, these being so great as to prohibit its frequent use in design problems. It also does not allow for the iron loss in the magnetic circuit except as the resistances may be arbitrarily increased, as was done in Case III above. There is little justification for an increase by the same factor in primary and secondary, although this is probably the most reasonable assumption to make.

III. SINGLE-COIL MODEL

As a contrast to the complex double-coil model which has just been discussed we may next consider the properties of the very simple circuit shown in Fig. 10, which may be called a single-coil model. If the coil in such a circuit is supplied with current from an external circuit which is suddenly interrupted at $t=0$, the current will at the start flow into the condenser, charging it to a continuously increasing voltage until finally the entire energy of the system is stored electrostatically in the condenser. After this the current will reverse and the circuit will oscillate in a manner analogous to either of the two circuits which constitute the double-coil model.

1. EQUATIONS OF MODEL

The differential equation for this simple type of circuit is

$$LC\frac{d^2v}{dt^2}+RC\frac{dv}{dt}+v-E=0 \tag{20}$$

where E is any constant emf in the circuit, such as that generated by rotation.

The solution of equation (20) to meet the initial conditions of condenser voltage equal to E and condenser current $C\frac{dv}{dt}$ equal to I_0 at the instant $t=o$ is:

$$v=\frac{I_0}{C(\alpha_1-\alpha_2)}(e^{\alpha_1 t}-e^{\alpha_2 t})+E \tag{21}$$

or

$$v=\frac{I_0}{\beta C}e^{\alpha t}\sin\beta t+E \tag{22}$$

where

$$\left.\begin{array}{ll} I_0=\text{current at break} & \alpha_1=-b+\sqrt{b^2-d^2} \\ b=\frac{R}{2L}=-\alpha & \alpha_2=-b-\sqrt{b^2-d^2} \\ d=\frac{1}{\sqrt{LC}} & \beta=\sqrt{d^2-b^2} \end{array}\right\} \tag{23}$$

Equation (21) applies to cases where $b^2>d^2$ and equation (22) to the more usual cases where $b^2<d^2$. The maximum value attained by the voltage in the former case is

$$V_m=\frac{-I_0}{C\alpha_2}y^{\frac{y}{1-y}}+E \tag{24}$$

where

$$y=\frac{\alpha_1}{\alpha_2}$$

or if the damping is so very great that d is negligible compared to b, equation (21) reduces to

$$v=\frac{I_0L}{RC}\left(e^{-\frac{t}{CR}}-e^{-\frac{Rt}{L}}\right)+E \qquad (25)$$

and eq (24) becomes in the limit when $\frac{R^2C}{L}$ approaches ∞

$$V_m=\frac{I_0L}{RC}+E \qquad (26)$$

Similarly, in the oscillatory case ($b^2<d^2$), the maximum voltage is

$$V_m=\frac{I_0}{dC}e^{\frac{\alpha}{\beta}\tan^{-1}\frac{\beta}{\alpha}}+E \qquad (27)$$

If the damping is very small, equations (22) and (27) reduce to

$$v=I_0\sqrt{\frac{L}{C}}e^{-\frac{Rt}{2L}}\ \sin\frac{t}{\sqrt{LC}}+E \qquad (28)$$

and

$$V_m=I_0\sqrt{\frac{L}{C}}+E \qquad (29)$$

The variation of voltage with time indicated by equation (22) has, of course, the damped sine wave form such as those plotted in Figs. 13, 14, and Curve II of Fig. 31, and it will be noted that these figures are similar to Fig. 7, except that the higher frequency ripple is absent. It is, therefore, apparent that by suitable choice of inductance, resistance, and capacity the simple model would give in its essential features the same performance as the more complicated circuits.

It will be noted that in equations (24), (26), (27), and (29) the maximum voltage attained, V_m, appears explicitly as proportional to I_0 the current at break, if the constant term E is neglected, and it is also evident that a similar proportionality exists in the other types of model. (See, for example, equation 67.) The ratio $\frac{V_m}{I_0}$ is consequently a convenient quantity for expressing the performance of the model. Since this quantity has the physical dimensions of an impedance, it will be termed the "impulsive impedance" of the system and denoted by Z in the following discussion.

2. DERIVATION OF SINGLE-COIL MODEL FROM ACTUAL CIRCUIT

The use of this single-coil model is closely analogous to the use of the "equivalent T circuit" which is so generally and usefully applied to alternating current problems in connection with transformer design and telephone circuits. The relation between the single-coil model and the double-coil model is perhaps best brought out by considering the series of steps by which one can pass physically from one to the other. For the connection shown in Fig. 1 one may imagine the space to the right of the dotted line replaced by an imaginary region in which the various electrical quantities are indicated by "primed" letters which are related to the actually existing electrical constants by the relations

$$R_2'=\frac{R_2}{p^2},\ L_2'=\frac{L_2}{p^2},\ C_2'=p^2C_2,\ I_2'=pi_2,\ v_2'=\frac{v_2}{p},\ E_2'=\frac{E_2}{p}, \qquad (30)$$

where p is any constant factor. In this fictitious space all the fundamental electromagnetic laws will still apply. For example,

$$i_2'=\frac{v_2'}{R_2'}\ \text{(Ohm's Law)}$$

$$E_2'=-L_2'\frac{di_2'}{dt},\ \text{etc.}$$

Furthermore, if the new secondary space is coupled to the original primary circuit by a value of mutual inductance given by

$$M'=\frac{M}{p} \qquad (31)$$

the reaction of any electromagnetic phenomenon on the right upon the circuits on the left will be the same as in the original space. If the ratio p is arbitrarily chosen as equal to the ratio of secondary to primary turns, the above procedure becomes what is called in electrical engineering "referring to the primary side" and applies not only to the constants of the particular transforming device, whether it be transformer, magneto, or induction coil, but also to the constants of any other apparatus connected beyond its terminals. Thus, if V_s (=5000 volts) is the actual breakdown voltage of a spark gap connected to the secondary, and if $p=50$, then

$$V_s'=\frac{V_s}{p}=100\ \text{volts}$$

is the breakdown voltage of the spark gap referred to the primary side.

This step has been purely arbitrary and mathematical, but we must now introduce the physical assumption that the primary and secondary coils are closely coupled, so that $M^2 = L_1L_2$ or $k = 1$. If we choose $p = \frac{M}{L_1} = \frac{L_2}{M}$ then

$$L_2' = M' = L_1, \tag{32}$$

and we may then make the next step and replace the two closely coupled coils by a single coil of inductance $L = L_1 = M' = L_2'$. The only physical phenomena affected by this last change are those involving the insulation between the coils and are of no concern

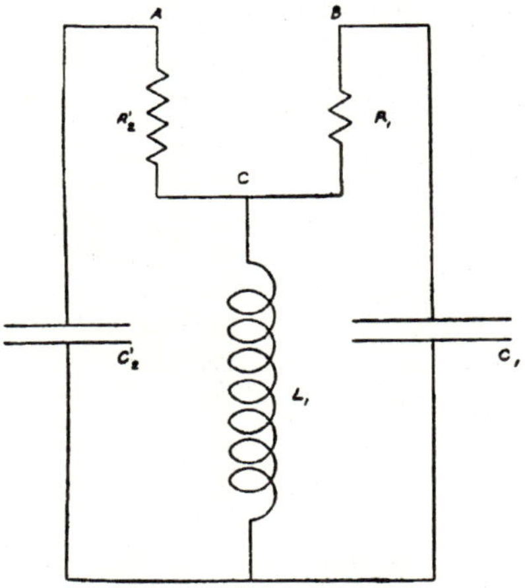

FIG. 11.—*Circuit intermediate between double-coil and single-coil models*

in the problem at hand. The result of the steps previously discussed gives a circuit shown in Fig. 11. If in any such network a current is established in the inductance coil L and left to itself, it will flow through the resistances R_1 and R_2', charging condensers C_1 and C_2'. If the relation

$$R_1C_1 = R_2'\,C_2' = R_2C_2 \tag{33}$$

holds, then the potentials of points A and B will rise together and be equal at all times like the galvanometer terminals of a Wheatstone Bridge, so that conditions will not be in any way changed if a connection is made between A and B. With such a connection in place, the circuits then become identical with those shown in Fig. 10 where $C = C_1 + C_2'$ and

$$R = \frac{R_1 R_2'}{R_1 + R_2'} \tag{34}$$

In any case the voltage established between C and either A or B is small compared to the voltage across the condenser or the coil, and the performance of the circuit shown in Fig. 11 is but little affected in case the relation (33) is not exactly fulfilled.

The justification of the single-coil model from a more mathematical point of view can be based upon equation (10) by inserting the primed quantities as defined by equation (30).

This substitution gives

$$C_1C_2'(L_1L_2' - M'^2)\frac{d^4v_2'}{dt^4} + C_1C_2'(L_2'R_1 + L_1R_2')\frac{d^3v_2'}{dt^3} + (R_1R_2'C_1C_2' + L_1C_1 + L_2'C_2')\frac{d^2v_2'}{dt^2} + (R_1C_1 + R_2'C_2')\frac{dv_2'}{dt} + v_2' = 0 \tag{35}$$

This shows that no modification of the differential equation results from the substitution and that the quantity v_2' is affected only as the insertion of the primed quantities has affected the limits of integration. As above, if $k = 1$, $(L_1 L_2' = M'^2)$ the first term of equation (35) vanishes and we are left with the third order equation

$$L_1C_1C_2' (R_1 + R_2') \frac{d^3v_2'}{dt^3} + \{R_1R_2'C_1C_2' + L_1(C_1 + C_2')\} \frac{d^2v_2'}{dt^2} + (R_1C_1 + R_2'C_2') \frac{dv_2'}{dt} + v_2' = 0. \tag{36}$$

Introducing the further physical assumption that

$$R_1C_1 = R_2C_2 = R_2'C_2' \tag{33}$$

we obtain

$$L_1R_1C_1 (C_1 + C_2') \frac{d^3v_2'}{dt^3} + \{R_1^2C_1^2 + L_1 (C_1 + C_2')\} \frac{d^2v_2'}{dt^2} + 2R_1C_1 \frac{dv_2'}{dt} + v_2' = 0 \tag{37}$$

which can be factored, giving

$$\left(R_1C_1 \frac{d}{dt} + 1\right) \left\{L_1 (C_1 + C_2') \frac{d^2}{dt^2} + R_1C_1 \frac{d}{dt} + 1\right\} v_2' = 0 \tag{38}$$

A complete solution of this is

$$v_2 = Ae^{-R_1C_1t} + Be^{\alpha t}\sin(\beta t + \theta) \qquad (39)$$
$$= X + Y$$

where A, B, and θ are constants of integration, and α and β are given by eq. (23) above, if we put

$$R = \frac{R_1 R_2'}{R_1 + R_2'} \text{ and } C = C_1 + C_2' \qquad (40)$$

Solution X represents currents flowing from one condenser into the other without magnetizing the coil (L does not enter), while Y is the same solution as that for the single-coil model. If both condensers are initially uncharged and there is an initial current I_0 in the coil, then the constant A of equation (39) becomes zero, and consequently the single-coil model gives a complete solution.

A similar line of reasoning on either the physical or mathematical basis may be applied to the type of connection shown in Fig. 2 and leads in a similar way to the single-coil model.

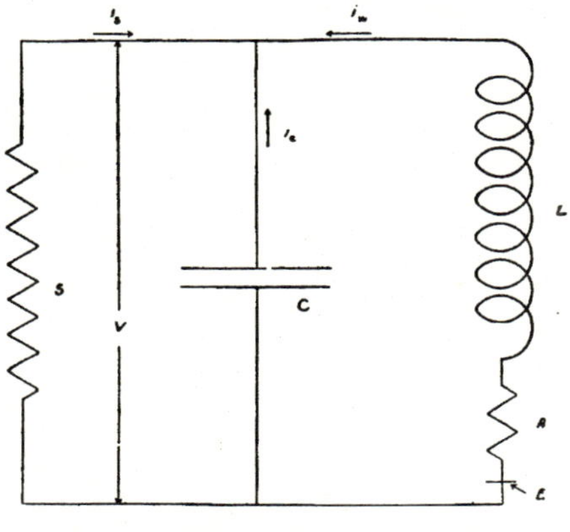

FIG. 12.—*Single-coil model with shunting resistance*

3. SINGLE-COIL MODEL WITH SHUNTING RESISTANCE

The case in which the secondary terminals are shunted by a resistance can be very readily handled by this type of model by using the substitution suggested by Mizuno.[15]

The circuit for such a case is shown in Fig. 12, and by Kirchhoff's laws we have, taking v as positive when the upper condenser plate is charged positively,

[15] F. Mizuno, Phil. Mag., **45**, p. 447; 1898.

$$i_w + i_s + i_c = 0 \tag{41}$$

$$v = -i_w R - L\frac{di_w}{dt} + E \tag{42}$$

$$i_c = -C\frac{dv}{dt} \tag{43}$$

$$i_s = -\frac{v}{S} \tag{44}$$

which may be combined to give

$$\frac{d^2v}{dt^2} + \left(\frac{R}{L} + \frac{1}{SC}\right)\frac{dv}{dt} + \left(\frac{1}{LC} + \frac{R}{LSC}\right)v - \frac{E}{LC} = 0 \tag{45}$$

or

$$L_1'C'\frac{d^2v}{dt^2} + R'C'\frac{dv}{dt} + v - E' = 0 \tag{46}$$

where

$$L' = L,\ R' = R + \frac{L}{SC},\ C' = C\frac{S}{R+S},\ E' = E\frac{S}{R+S} \tag{47}$$

It will be seen that equation (46) is identical in form with equation (20) above, so that the single-coil model when shunted is equivalent to an unshunted coil having the fictitious resistance and capacity given by equation (47). It must be noted, however, that the initial conditions of the shunted model are at $t=0$.

$$i_w = I_0 \text{ and } v = E \tag{48}$$

and are not the relations

$$C'\frac{dv}{dt} = I_0 \text{ and } v = E' \tag{49}$$

which would be obtained by strictly following through the analogy. Solving the above equations gives, in case the circuit is overdamped,

$$v = \frac{I_0}{C(\alpha_1' - \alpha_2')}\left(e^{\alpha_1' t} - e^{\alpha_2' t}\right) + E'\left[\frac{S}{R+S} + \left(\frac{1}{SC} + \frac{R\alpha_1'}{R+S}\right)\frac{e^{\alpha_2' t}}{\alpha_1' - \alpha_2'} - \left(\frac{1}{SC} + \frac{R\alpha_2'}{R+S}\right)\frac{e^{\alpha_1' t}}{\alpha_1' - \alpha_2'}\right] \tag{50}$$

or, if E is negligible,

$$v = \frac{I_0}{C(\alpha_1' - \alpha_2')}\left(e^{\alpha_1' t} - e^{\alpha_2' t}\right) \tag{51}$$

where α_1' and α_2' are obtained from L', R', and C' by equations (23). It is interesting to note that when S is so small as to produce very heavy overdamping, and make d^2 entirely negligible compared with b^2, we have the simple relation

$$V_m = I_0 S \tag{52}$$

This relation, of course, might have been obtained directly from the physical consideration that the coil will tend to send its full current I_0 through the shunting resistance S and will consequently establish across the terminals of the condenser a voltage equal to the IR drop $= I_0 S$. In most ignition circuits which are so heavily fouled as to cause misfiring this condition is quite closely approached, and a knowledge of the shunting resistance S and the sparking voltage of the gap referred to the primary side immediately gives, by comparison with the primary current at break I_0, an indication whether or not a spark will be obtained.

In case the shunting resistance is not so low as to produce excessive damping and if E is negligible, the solution of equation (46) is

$$v = \frac{I_0}{C\beta'} e^{\alpha' t} \sin \beta' t \tag{53}$$

where, as before, α' and β' are obtained from R', L', and C' by equation (23).

4. NUMERICAL EXAMPLES

From equations (21), (22), (51), and (53) above there may be computed, as shown by Figs. 13, 14, and 15, the variations of voltage, current, and of energy with time under various conditions. Case I corresponds to a circuit having $L = 0.015$ henry, $C = 0.32$ microfarad, $R = 16.8$ ohms. This circuit is that which would be obtained from that considered as Case III of the double-coil model on referring all the constants to the primary side. Case II (Fig. 13) corresponds to the same circuit when it is shunted with an additional capacity of 100 $\mu\mu f$ on the secondary side or 0.24 microfarad on the primary side. Such a shunting might well be produced by the addition of leads in a metal tube from the magneto terminal to the spark plug, or by a reasonable increase of primary condenser, such as might be thought necessary to reduce sparking at the break. It is seen that the addition of this capacity increases the period of oscillation and the time interval between the breaking of the primary current and the attainment of maximum voltage. It also produces a considerable reduction in the magnitude of the maximum which is obtained.

Cases III and IV correspond to the same circuit treated in Case I, but shunted in Case III by a resistance of 500 000 ohms on the secondary and in Case IV by 100 000 ohms on the secondary. The latter shunting is sufficient to make the oscillations aperiodic. Fig. 14 shows the very marked decrease in maximum voltage which results from the presence of the shunting resistance and Fig. 15, which shows the fluctuations in energy for Cases I, III, and IV, indicates the character of the energy transformations which take place.

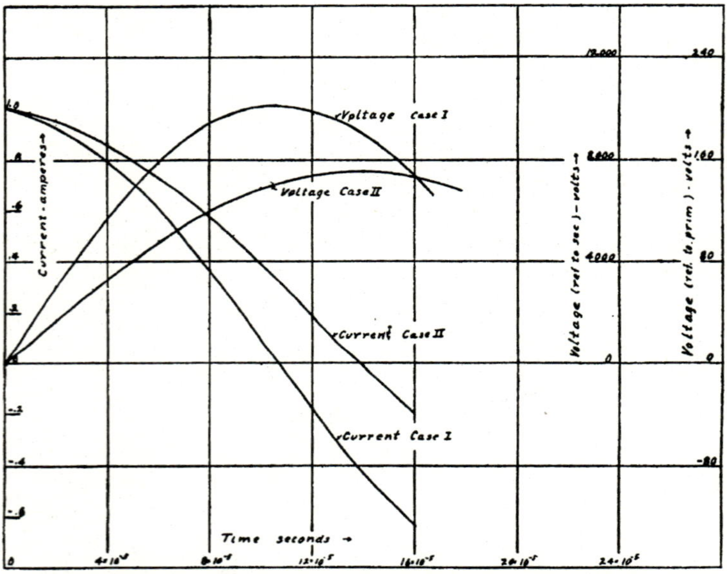

Fig. 13.—*Effect of added capacity on voltage and current wave forms during period 2 computed for single-coil model*

Case I has $L=0.015\ h$, $C=0.32\ \mu f$, $R=16.8\ \Omega$, $n=49$. Case II is for the same model with 100 micro-microfarads additional capacity in parallel with the secondary

In Case I the electrostatic energy in the condenser rises to a maximum at $t=11\times10^{-5}$ seconds, at which time all of the energy then present exists in the condenser. The drop in total energy from the initial value of 7.5×10^{-3} joules results from the dissipation of energy in the resistance. In Case III, however, we have at time $t=8\times10^{-5}$ seconds a maximum storage of energy in the condenser and a maximum voltage across its terminals. At this time the total energy has, as a result of the dissipation of energy in the shunting resistance, fallen to 4×10^{-3} joules, and the remaining 2×10^{-3} joules are still stored as magnetic energy in the coil and are maintained in that form by the current flowing through the coil and the shunting resistance. At the later instant $t = 17 \times 10^{-5}$ seconds the current in the coil is passing through zero, but the entire energy supply has been so much depleted that at this time there is less in the condenser than at the earlier instant. Still later in the cycle the current builds up in the reverse direction and receives some of the slender store of energy which is still flowing out of the condenser. The storage

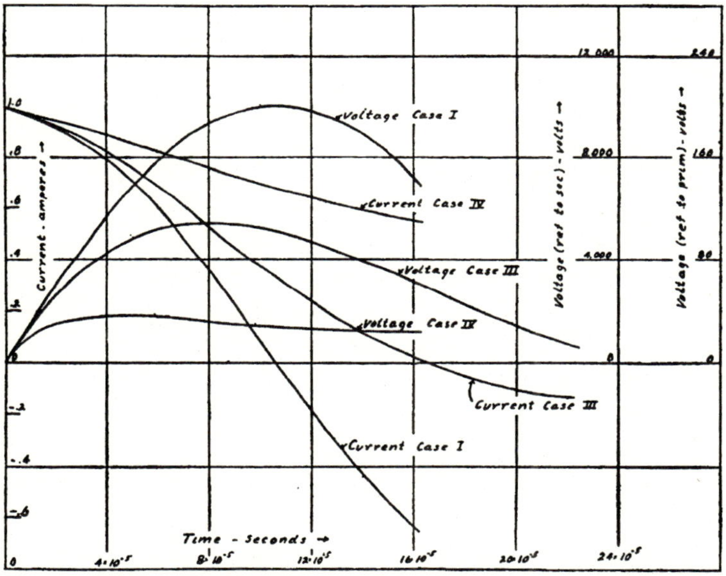

FIG. 14.—*Effect of shunting resistance on voltage and current wave forms during period 2 computed for single-coil model. Case I has L=0.015 h, C=0.32μf, R=16.8 Ω, n=49. Case III has in addition a shunting resistance of 500 000 Ω on the secondary (equivalent to 208 Ω when referred to the primary). Case IV has a shunting resistance of 100 000 Ω on the secondary (equivalent to 41.7 Ω when referred to the primary)*

of magnetic energy in the coil by the current through the shunting resistance is closely analogous with the effect which will be discussed later of the storage of energy by the eddy currents in the core of the magneto at the instant when the currents in the windings are zero. This similarity also brings out the possibility of expressing the eddy current loss in terms of an equivalent shunting resistance and using Fig. 12 as a type of model to express these eddy current effects. It is probable, however, that in most quantitative work the alternative circuit shown in Fig. 10

with an increased value of R will be found more useful, and for a precise study of these effects the more complex "closed-coil model," to be discussed below, must be resorted to.

In most cases it is only the maximum voltage of the first oscillation which is of importance and which is given by equations (24), (26), (27), and (29), and the effect of such quantities as capacity and shunting resistance on the voltage may be represented directly by curves such as those given in Figs. 32 and 33 below, in which these physical values are compared with experi-

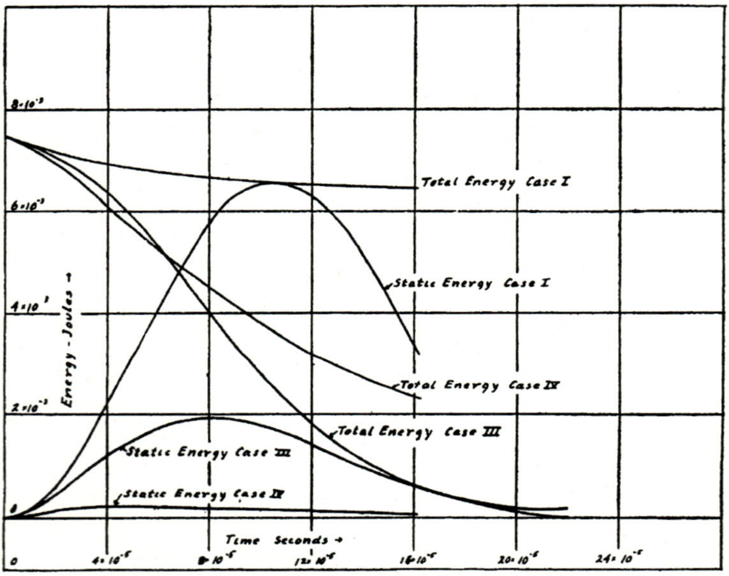

FIG. 15.—*Changes of energy with time computed for single-coil model with various shunting resistances*

mental results. The general characteristics of Fig. 32 are, of course, determined by the fact that the initial amount of energy, $\frac{1}{2} L I_0^2$, is later in the cycle stored with relatively little loss in the condenser C, and the resulting voltage may be computed from the relation

$$\tfrac{1}{2} L I_0^2 = \tfrac{1}{2} C V_m^2 \text{ or } V_m = I_0\sqrt{\frac{L}{C}} \qquad (54)$$

The general characteristics of the curve shown in Fig. 33 are that for very high values of the shunting resistance the maximum voltage approaches a constant value given by the preceding relation, while with very low values of the shunting resistance the simple linear relation of equation 52 applies. The dotted line shows the relation of impulsive impedance to shunting resistance on this basis.

5. APPLICATION OF SINGLE-COIL MODEL TO COMPLETE CYCLE

The relatively great simplicity of the single-coil model as compared with either the double-coil model or the closed-coil model which is discussed below renders it quite suitable for computation, and it has the further advantage of lending itself with very slight extension to a qualitative indication of the complete cycle of

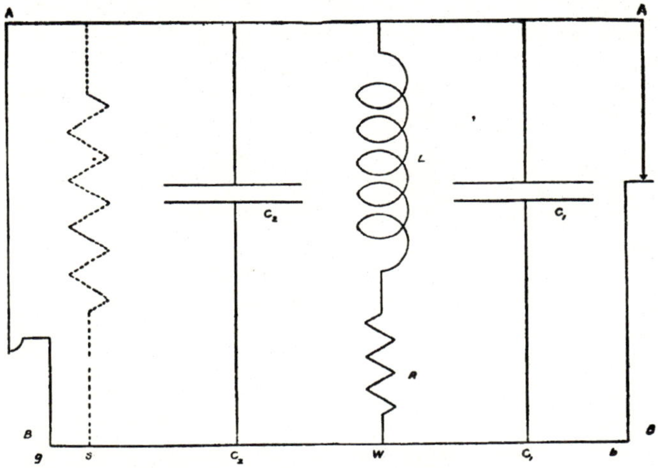

FIG. 16.—*Single-coil model extended to indicate complete cycle of operation of magneto*

operation of the magneto. Fig. 16 shows the single-coil model as thus extended by the addition of components corresponding to the primary circuit breaker and the spark gap and with its total capacity again divided into the components C_1 and C_2, of which it was originally composed.

It is seen that this model consists approximately at least of two perfect conductors or junction points *A–A* and *B–B* between which are or may be connected five circuits in parallel. At the beginning of the cycle of operation, during what may be called period 1, we have contact breaker *b* closed and circuits *w* and *b* of importance in the phenomenon. During this period the rotation of the armature between the magnet poles generates a current in the circuit thus formed, which, at the instant of break, has attained a value, I_0. At the instant of break conditions are abruptly changed by the elimination of circuit *b* from the system and we then have, as has been discussed above, the coil *w*, charging the two condensers C_1 and C_2 in parallel at an initial rate I_0. This continues until at some short time later the voltage across the condenser C_2 and across the spark gap *g* has risen to a value equal to the breakdown voltage of the gap. In most cases this occurs before the current I_0 has decreased very greatly and before more than a small percentage of the total magnetic energy has been stored in the two condensers. When the spark gap *g* becomes conducting, which occurs with great rapidity, we have for the very short period 3 the discharge of the two condensers through the spark gap, so that the circuit *g* in series with the combination of C_2 and C_1 in parallel only need be considered. Actually the leakage reactance of the winding, though, as explained above, nearly negligible during period 2, is of fundamental importance during period 3 because of the very high frequency of the oscillations of the circuit $g\ C_2\ C_1$. Consequently, we may expect that only condenser C_2 will be discharged at first through spark gap *g*, and that the discharge of condenser C_1 will follow at an appreciably later time after it has established the magnetic field required by the leakage flux of the winding. At the end of this entire period 3, which is of extremely short duration, we are left with the coil *w* in series with spark gap *g*, which is now broken down and which may be considered as replaced by a relatively high resistance (10 000 to 40 000 ohms on the secondary or 4 to 16 ohms referred to the primary) through which the coil can discharge. This discharge takes place during what may be called period 4 and continues usually until the entire store of energy is dissipated.

This interval of relatively long duration (0.001 to 0.008 second) is what is recorded by an oscillograph connected in series with the spark gap. If the entire supply of energy is dissipated before the contact *b* is closed, there is no further action until period 1 of the next cycle begins. If, however, the cam is so short or the magneto speed so high that the circuit breaker *b* closes before the exhaustion of the energy supply, we then have circuit *b* of relatively low resistance in parallel with circuit *g* of high resistance, both being fed by the coil *w*. The result is that most of the current flows through circuit *b* and, owing to the curious volt-ampere characteristic of spark gaps, according to which their resistance increases rapidly as the current flow through them decreases, the entire current is actually diverted from circuit *g* to circuit *b* and the spark is extinguished with great suddenness. Furthermore the condensers which had previously been charged to the relatively low voltage (say, 1200 on secondary or 24 referred to primary) which was required to maintain the current through the spark gap now discharge themselves rapidly and probably in an oscillatory manner through circuit *b* during a short interval which may be called period 5. After the cessation of this condenser discharge the coil may still continue to send considerable current through circuit *b* during what may be called period 6 until the energy supply is finally exhausted. It will be found that this conception of a single-coil model as drawn in Fig. 16 is of great utility in analyzing the results of oscillograph tests of magnetos and indicates the reasons for the quantitative relations between corresponding changes in primary and secondary currents resulting from the opening and closing of the breaker or fluctuations in circuit resistance. It is obvious that the addition of a sixth circuit, *s*, containing resistance, as indicated by the dotted lines, serves to take account of the presence of shunting resistance across the spark plug points.

IV. CLOSED-COIL MODEL

1. EQUATIONS

The two models previously discussed can account for the energy losses in the iron of the magnetic circuit only by the device of increasing the resistance of one or both coils by some arbitrary amount. This is theoretically unsatisfactory, since it essentially implies that the energy loss is proportional to the square of the current flowing in the winding, which in general may not be correct. The iron loss consists of a certain amount of hysteresis loss, together with the dissipation of energy which results from eddy currents induced in the conducting iron core as a result of the rapid rate of change of flux. In the normal operation of the magneto the magnetic change which occurs during period 2 consists of a sudden decrease of the magnetic flux, and the energy which is lost as a consequence of this process can be estimated from the departure of the actual flux-magnetomotive force curve of the magnetic circuit from the ideal straight line which would correspond to a circuit having no such loss. Measurements of the entire hysteresis loss for circuits of this type[16] indicate that the total amount of hysteresis loss is small relative to the energy stored in the circuit, and consequently may be neglected without introducing serious error. The larger portion of the iron loss is due to eddy currents, and consequently their effect upon the operation of the device may, to a first approximation, be represented by the presence of a closed tertiary winding which is coupled inductively to the actual physical winding of the core and which possesses certain resistance and inductance as well as a

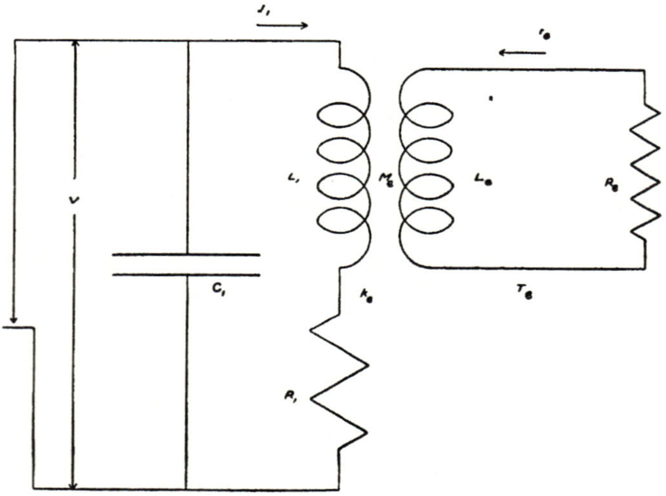

FIG. 17.—*Circuits of closed-coil model of magneto*

certain mutual inductance to the winding. The addition of such a tertiary circuit to the double-coil model would lead to an aggregation of circuits which would be too complex for useful application, and it is more valuable to add the closed tertiary circuit to the single-coil model just discussed, thus obtaining a circuit shown in Fig. 17, which will be called the "closed-coil model." A little consideration will show that the performance of the external winding will not be affected if, in place of any particular assumed tertiary circuit, an alternative circuit is substituted in which

$$L_e' = p^2 L_e, \quad M_e' = p M_e, \quad R_e' = p^2 R_e. \tag{55}$$

[16] A. P. P. Report No. 20, Nat. Advisory Comm. for Aero. Report No. 58:: 1919.

It consequently appears that the measurements and effects in the actual winding depend only upon two constants of the tertiary coil and not upon the three constants L_e, R_e, M_e, as would appear at first sight. For these two constants the most convenient ones to use are the coefficient of coupling k_e, defined as

$$k_e=\sqrt{\frac{M_e^2}{L_1L_e}} \tag{56}$$

and the time constant T_e of the circuit defined as

$$T_e=\frac{L_e}{R_e} \tag{57}$$

Applying Kirchhoff's laws to the circuit shown in Fig. 17 gives

$$L_e\frac{di_e}{dt}+R_ei_e+M_e\frac{di_1}{dt}=0 \tag{58}$$

$$L_1C_1\frac{d^2v}{dt^2}+M_e\frac{di_e}{dt}+R_1C_1\frac{dv}{dt}+v=E_1 \tag{59}$$

which may be combined with eqs. (56) and (57) to give

$$T_e(1-k_e^2)L_1C_1\frac{d^3v}{dt_3}+\left(1+\frac{R_1T_e}{L_1}\right)L_1C_1\frac{d^2v}{dt^2}+\left(1+\frac{T_e}{R_1C_1}\right)R_1C_1\frac{dv}{dt}+v-E_1=0 \tag{60}$$

E_1 may usually be neglected, or if its derivatives are negligible it may be eliminated by a change of variable, as was done in equation (7) above. On either basis the solution of equation (60) depends upon that of the corresponding auxiliary equation.

$$T_e(1-k_e^2)L_1C_1m^3+\left(1+\frac{R_1T_e}{L_1}\right)L_1C_1m^2+\left(1+\frac{T_e}{R_1C_1}\right)R_1C_1m+1=0 \tag{61}$$

The initial conditions are $v=E_1$, $i_1=I_0$ and $i_e=0$ at $t=0$, and two cases arise, depending on whether the three roots of (61), m_1, m_2, and m_3 are all real and negative, or whether m_1 is real and negative, while m_2 and m_3 form a pair of conjugate complex quantities, so that

$$m_2=\alpha+j\beta;\ m_3=\alpha-j\beta \tag{62}$$

In the former case the solution of (60) is

$$v = A_1 e^{m_1 t} + A_2 e^{m_2 t} + A_3 e^{m_3 t} \tag{63}$$

where

$$\left.\begin{aligned} A_1 &= \frac{\left(m_2 + m_3 + \frac{R_1}{(1-k_e^2)L_1}\right)I_0}{(m_1 - m_2)\,(m_3 - m_1)C_1} \\ A_2 &= \frac{\left(m_3 + m_1 + \frac{R_1}{(1-k_e^2)L_1}\right)I_0}{(m_2 - m_3)\,(m_1 - m_2)C_1} \\ A_3 &= \frac{\left(m_1 + m_2 + \frac{R_1}{(1-k_e^2)L_1}\right)I_0}{(m_3 - m_1)\,(m_2 - m_3)C_1} \end{aligned}\right\} \tag{64}$$

In case two of the roots (say m_1 and m_2) are nearly equal the corresponding A's become very large and equations (63) and (64) become inconvenient. If we set $m_2 = m_1 + \delta$ (65)

equation (63) becomes (assuming $\frac{R_1}{(1-k_e^2)L_1}$ to be negligible)

$$v = A_3(e^{m_3 t} - e^{m_1 t}) + \delta\, A_2 t e^{m_1 t} \quad . \; . \; . \; . \text{ approx.} \tag{66}$$

if δt is small compared to 1.

When m_2 and m_3 are complex, the solution can be put in the form

$$v = I_0\left[\frac{e^{\alpha t}}{C_1\beta}\sqrt{\frac{(\alpha+m_1+A)^2+\beta^2}{(\alpha-m_1)^2+\beta^2}}\sin(\beta t-\varphi) - \frac{2\alpha+A}{C_1\{(\alpha-m_1)^2+\beta^2\}}e^{m_1 t}\right] \tag{67}$$

where α and β are defined by equation (62) and

$$A = \frac{R_1}{L_1(1-k_e^2)} \tag{68}$$

$$\varphi = \tan^{-1}\frac{\beta(A+2\alpha)}{\alpha^2 - m_1^2 - \beta^2 + A(\alpha - m_1)} \tag{69}$$

In the practical use of the closed-coil model it is seldom necessary to carry out a direct solution of the cubic equation (61), as it is usually easier to obtain one root m_1 by successive approximations. The root m_1 has the value $-\frac{1}{T_e}$ for large values of T_e and the value $\frac{1}{T_e(1-k_e^2)}$ for small values of T_e, so that an assumed value between these limits gives a good first approximation. Equation (61) may be transposed to give

$$\frac{1}{T_e} = -\left(m - \frac{k_e^2 L_1 C_1 m^3}{L_1 C_1 m^2 + R_1 C_1 m + 1}\right) \tag{70}$$

and several assumed values of m_1 can be substituted in the right-hand member and the corresponding value of T_e computed. An auxiliary graph can then be drawn with T_e as ordinate against m_1 as abscissa, and the precise value of m_1 corresponding to the original value of T_e is then read off. The cubic equation can then be depressed by dividing through by the expression $(m-m_1)$ and the resulting quadratic be easily solved by the usual methods.

It is impracticable to obtain an explicit expression for the maximum voltage of the impulse defined by equations (66) or (67), as

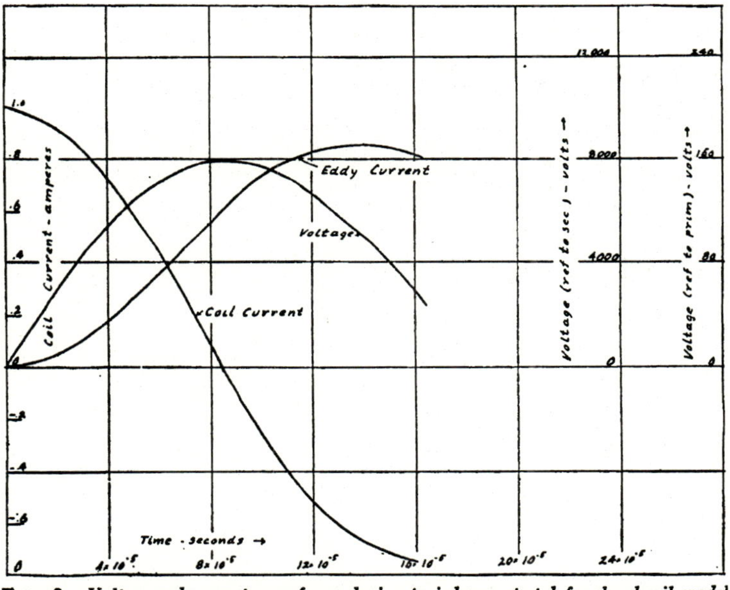

Fig. 18.—*Voltage and current wave forms during period 2 computed for closed-coil model having* $R_1=0.33\,\Omega$, $L_1=0.015$ *h*, $C_1=0.325\,\mu f$, $k_e^2=0.5$, $T_e=1.80\times10^{-4}$ *sec.*

was done in the case of the single-coil model. In a given numerical case the voltage may be computed for several values of t near the crest of the impulse and the maximum value read from an auxiliary graph of V versus t. A value of

$$t=\frac{\frac{\pi}{2}+\varphi}{\beta} \tag{71}$$

is a convenient first approximation, but is always slightly too large.

2. NUMERICAL EXAMPLE

Fig. 18 and Curve III of Fig. 31 show the general character of the variation of voltage and current in the various circuits with time and Fig. 19 the corresponding energy changes. It will be noted that the term involving e^{m_1t} (shown dotted in Fig. 31) in effect shifts the zero line with respect to which the damped oscil-

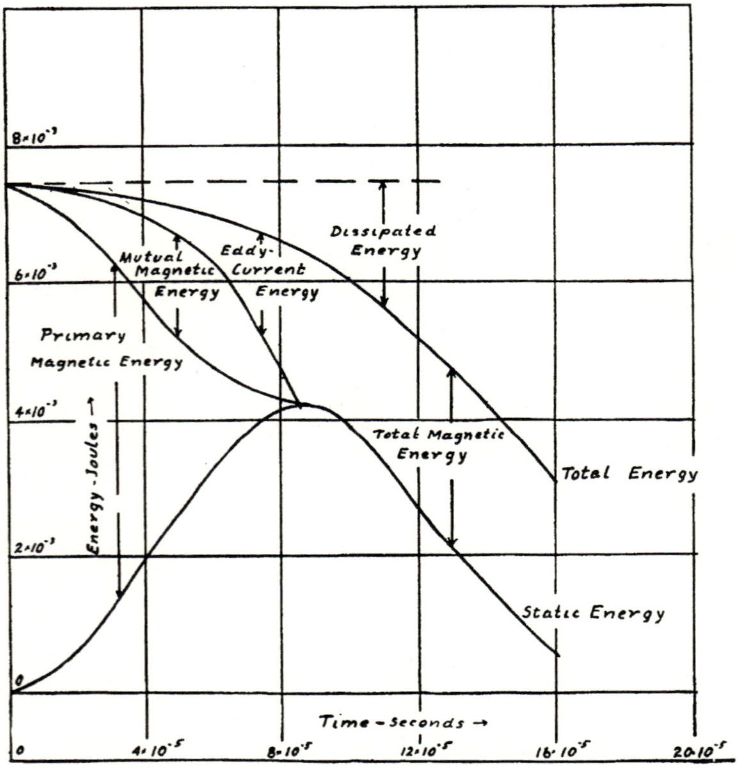

FIG. 19.—*Changes of energy with time. Closed-coil model*

lation represented by the second term occurs and gives a corresponding increase to the first positive maximum voltage and a slightly less decrease to the amplitude of the first negative maximum. This tendency is very clearly shown in the experimental curves obtained by Campbell, Fig. 9, and can not be produced by either the single or double coil models directly.

It will be noticed that in case the quantity T_e is very large relative to $\sqrt{L_1C_1}$ that the equation (61) may be factored to give

$$\left(m+\frac{1}{T_e}\right)\left\{(1-k_e^2)L_1C_1m^2+\left(R_1C_1+\frac{k_e^2L_1C_1}{T_e}\right)m+1\right\}=0 \qquad (72)$$

The difference between the approximate equation (72) and the correct equation (61) is a term $k^2_e\dfrac{L_1C_1}{T_e}m$, which becomes negligible in comparison with $T_e\, m$ for large values of T_e.

Similarly, if T_e is very small, the approximate equation

$$\left(m+\frac{1}{T_e(1-k_e^2)}\right)\{L_1C_1m^2+(R_1C_1+k_e^2T_e)m+1\}=0 \qquad (73)$$

may be used. The difference between equations (73) and (61) consists of the term $-k^2_e\{T_eR_1C_1+T^2_e(1-k^2_e)\}m^2$, which becomes negligible in comparison with $L_1C_1m^2$ for sufficiently small values of T_e.

The terms in the final solution corresponding to the first factor in equations (72) and (73) are negligible for the respective ranges of T_e for which the equations apply, so the resulting solutions in these two extreme cases are in the same form as those given by the single-coil model. It is therefore evident that the substitution of proper values of L and R in the latter model will give the same results as the more complicated equation (67) above for either extreme case. It so happens, however, that in a typical shuttle armature magneto experimented with the observed values of T_e were of the same order of magnitude as $\sqrt{L_1C_1}$, and consequently the simplified equations were not applicable.

The decrease in the coefficient of m^2 in equation (72) as compared with the analogous term in equation (73) or equation (20) can be roughly explained physically if we assume that the coupling between the tertiary coil and the main coil is direct and amounts to shunting a fraction $\dfrac{1}{k_e}$ of the total winding by a resistance R_e. If the time constant T_e is made very large and consequently R_e made very small, the net effect is to short circuit $\dfrac{1}{k_e}$ of the winding, leaving only a fraction of the original inductance remaining.

Fig. 20 shows the values of the real root m_1 in equation (61) for various values of T_e, and k^2_e in a model having $L_1=0.015$ *h*, $R_1=0.33$ ohm, $C_1=0.325$ μf, curves being drawn for values of $k^2_e=0.1$,

0.5, 0.9, and 0.95. It may be shown that all the roots of equation (61) will be real only if k^2_e is greater than 0.889.[17]

It is seen by Fig. 20 that for such values of k^2_e, the curves of m_1 against T_e have the double bend characteristic of the cubic equation, and the three real values of m correspond to intersections of this curve with an ordinate at the proper value of T_e. The principal effect of increasing T_e at any value of k_e is to decrease the maximum voltage from the value that it has with no eddy current present to a reduced value which becomes constant for

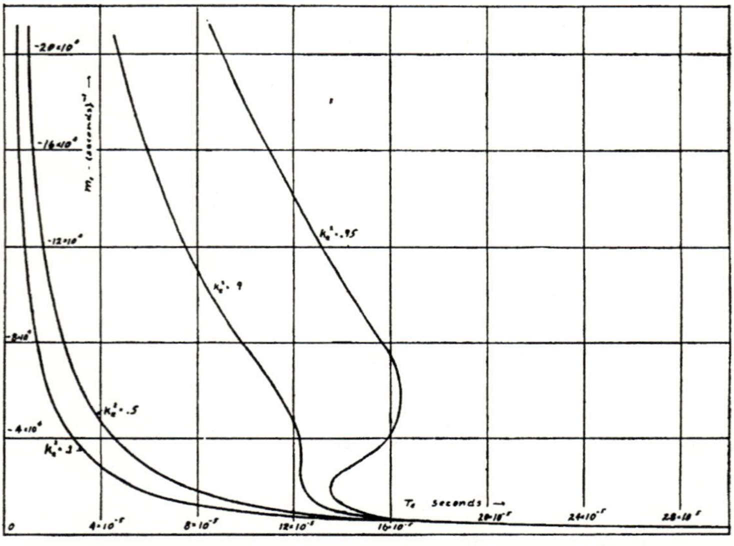

FIG. 20.—*Curves showing values of the real root m_1 of cubic equation (62) for various values of T_e and k_e^2 for circuits having R_1=0.33 Ω, L_1=0.015 h, C_1=0.325 μf*

large values of T_e. This is shown in Fig. 21, in which the crest voltage computed for a particular double-coil model (R=0.33 ohm, L=0.015h, C=0.325μf) is plotted against T_e for various values of k^2_e. The lower curves indicate the contribution to this total crest voltage of the term involving $e^{m_1 t}$ of equation (67), and hence the displacement of the axis about which the oscillation corresponding to the second term takes place. It will be noted that this effect is relatively small unless k^2_e is nearly unity and T_e is greater than the limiting value at which all three roots of equa-

[17] For a discussion of the frequencies and damping coefficients in this type of closed-coil model see Iolo-Jones, Phil. Mag., 39, p. 553, May, 1920.

tion (61) are real. For smaller values of T_e the e^{m_1t} term is negligibly small.

In general T_e will be proportional to the conductivity of the material, and hence will be decreased by the use of thinner laminations. It is indicated by the curve that no material change in the secondary voltage is to be expected unless the decrease in lamination thickness serves to make the value of T_e less than the value of $2\sqrt{L_1C_1}$.

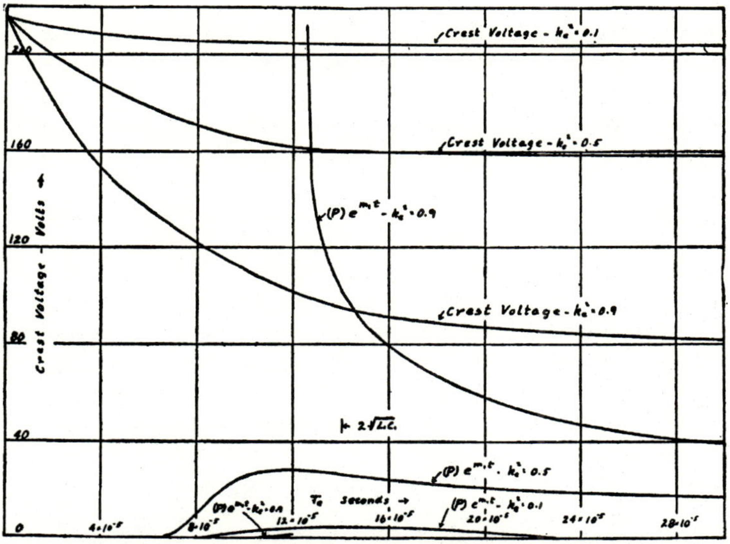

Fig. 21.—*Variation of crest voltage of single-coil model with T_e for various values of k_e^2. The lower curves show the contribution to the crest voltage of the last term* (P) e^{m_1t} *of Eq. 67*

3. CORRELATION WITH SKIN EFFECT IN IRON CORE

All of the above conclusions have been based on the assumption of a single eddy current circuit of fixed constants. It is possible, however, by the alternating current bridge described below to determine what constants the equivalent tertiary coil must have in order to account for the observed resistance and inductance of the main winding at any particular frequency at which the bridge measurements are made. The results of such measurements are shown in Fig. 29, which shows the variation of T_e and k^2_e, respectively, with the frequency used in the measurements. The two sets of curves correspond to magnetos which are identical except that in one the pole pieces are laminated, while in the other they are solid iron castings. It will be noted that the variation of T_e with frequency is very marked and that the quantity ωT_e is more nearly constant, although it increases with increase in frequency.

These variations of the tertiary coil constants with frequency are, of course, the result of the magnetic skin effect in the laminations and indicate that at higher frequencies the magnetic flux does not have time to penetrate throughout the thickness of the lamination, but that the flux density and also the eddy current density are greater near the surface than in the center. This spreading outward of the eddy current circuit accounts for the increase in the effective coupling k_e, and the resulting increase in the resistance offered to the flow of the eddy current due to the skin effect accounts for the marked increase in R_e and decrease in T_e.

A rough estimate of the amount of this skin effect to be expected can be obtained by considering the effective inductance and resistance of a uniformly wound coil containing a laminated iron core which has a relative short air gap. By making a number of simplifying assumptions, such as the absence of magnetic leakage, a uniform distribution of magnetic flux density over the cross section of the core on direct current and a permeability which is independent of the flux density we may deduce an expression for the effective impedance as being

$$Z = R + j\omega L_1 \left(\frac{h + \mu g}{sh + \mu g} \right) \tag{74}$$

where R = resistance of coil to direct current.

L_1 = inductance of coil to direct current.

μ = permeability.

s = ratio of the flux density at the surface of the iron to the mean flux density throughout the cross section. This is, in general, a complex quantity.

h = length of iron core.

g = length of air gap.

The distribution of eddy currents in an iron lamination has been worked out[18] and the relation connecting the flux density B_1 at the surface of the lamination, which is also equal to what would be produced throughout the material at zero frequency, with

[18] C. P. Steinmetz, Transient Electrical Phenomena and Oscillations, p. 365, 3d ed.

the mean flux density B_m throughout the thickness, which actually exists at the given frequency, is found to be

$$s=\frac{B_1}{B_m}=(1+j)\ cl_0\ \frac{\cosh c\ l_0 \cos c\ l_0 - j \sinh c\ l_0 \sin c\ l_0}{\sinh c\ l_0 \cos c\ l_0 - j \cosh c\ l_0 \sin c\ l_0} \qquad (75)$$

where $c=\sqrt{4\pi^2\frac{\mu}{\rho}f}$, ρ = resistivity, μ = permeability, f = frequency. (All in c g s units.)

For large values of cl_o, this reduces to

$$s=(1+j)\ cl_o \qquad (76)$$

Inserting this value of s in the equation (74) above leads to the relations

$$Z=R+j\omega L_1\left(\frac{1+\frac{\mu g}{h}}{(1+j)cl_o+\frac{\mu g}{h}}\right) \qquad (77)$$

or

$$Z=R+j\omega L_1\frac{1+x}{(1+j)cl_o+x}$$

if we let

$$x=\frac{\mu g}{h} \qquad (78)$$

Now, it may be shown that if alternating current of frequency f is passed through the primary of the closed coil model (Fig. 17) the apparent impedance of the coil is

$$Z=R_1+\frac{\omega^2 k_e^2 T_e L_1}{(1+\omega^2 T_e^2)}+j\omega L_1\left(\frac{1+\omega^2 T_e^2(1-k_e^2)}{1+\omega^2 T_e^2}\right) \qquad (79)$$

equating the two expressions (77) and (79) for the apparent impedance gives

$$k_e^2=\frac{cl_o(1+\omega^2 T_e^2)}{\omega T_e(cl_o\omega T_e+cl_o+x)} \qquad (80)$$

and

$$T_e=\frac{-(cl_o+x)(1-cl_o)+c^2l_o^2}{\omega cl_o(1+x)} \qquad (81)$$

as being the values of k_e and T_e to be expected for coils containing such a laminated iron core.

The results of equations (80) and (81) are plotted in Fig. 22, the scale of frequency being chosen to correspond to a value of $\frac{cl_o}{\sqrt{f}}$ equal 0.1, which, in turn, is a reasonable assumption in case the iron has a permeability $\mu = 1000$ and a resistivity $\rho = 1000$. These curves are plotted for values of x (the ratio of air gap reluctance divided by iron reluctance) of 2, 5, and 10, respectively. It will be noted that the curves connecting ωT_e with frequency run approximately parallel with those observed for one of the magnetos tested, and that an agreement could be obtained by a suitable choice of the factor x. The agreement between the curves connecting k_e^2 with frequency is not so good, and it is possible that a decidedly different value of $\frac{cl_o}{\sqrt{f}}$ would give a better agreement.

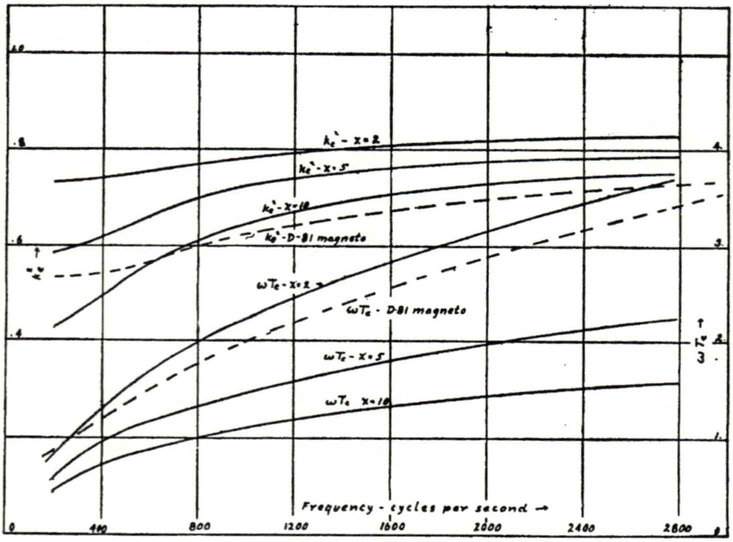

FIG. 22.—*Variation of k_e^2 and ωT_e with frequency*

Solid curves give values computed from skin effect equations (72 and 73) for three values of $x = \frac{\text{air-gap reluctance}}{\text{iron reluctance}}$ assuming uniform laminations and that $4\pi^2 \frac{\mu}{\rho} l_0^2 = 0.01$. Dotted curves give values observed on shuttle type magneto

The general order of magnitude and the rate of increase with frequency is the same in the theoretical and actual cases, and the discrepancy probably results from the presence of large and unlaminated masses of metal in the end cheeks of the armature, the magnet poles, the supporting framework, etc., no account of which is taken in the theoretical discussion.

The energy relations between the main circuit and the tertiary circuit are shown in Fig. 19, which is a plot of the electrostatic energy in the condenser, the total magnetic energy and the amount of energy dissipated as heat. It will be seen that at the time $t=9\times10^{-5}$ seconds, at which the voltage across the condenser is a maximum, only 1.1×10^{-3} joules, or 15 per cent, of the energy has been dissipated, but 2.1×10^{-3} joules, or 28 per cent, additional is rendered unavailable by the presence of the eddy currents and is stored magnetically in that part of the iron circuit which is linked with these currents. After the breakdown of the spark gap a considerable portion of this 28 per cent may be liberated and restored to the main circuit to later appear as heat in the spark, so that from this point of view measurements of spark energy do not entirely indicate the amount of energy available for the production of voltage. This plot also indicates the similarity between the closed-coil model and the single-coil model containing a shunting resistance in parallel with its terminals. The two cases would be identical if k_e^2 were unity and the qualitative agreement is fairly close in any case.

V. MEASUREMENTS OF CONSTANTS

In determining the electrical constants *R*, *L*, *C*, etc., to insert in the models previously discussed two general methods are possible: (1) By independent measurements of physical dimensions or electrical quantities, such as capacity of condensers, etc.; (2) by measuring the performance of the complete device and setting up a sufficient number of equations to solve empirically for the set of constants which will best satisfy the observed results. The former type of method is, of course, desirable from the point of view of the designer or testing laboratory, while only the second type of method is applicable in cases where it is desirable to study and express in quantitative terms the performance of apparatus already in existence.

Capacity Measurement

The capacity C_1 of the primary condenser of the magneto may, of course, be easily measured by any of the standard methods and need not be referred to here. The secondary distributed capacity consists of several parts, as indicated schematically in Fig. 23: First, the capacity C_t of the condenser, which has for one "plate" the high-tension lead and the outer layer of the secondary winding, and for the other "plate" the frame of the machine, the engine ground, and any tube through which the spark-plug cable may pass. Portions of this capacity may be isolated and measured separately, but the portion between the outer layer of the winding and the frame can not be handled in this manner, since even the removal of the armature from the frame merely substitutes the ends of the core and surrounding objects for the magneto frame, and while it materially reduces C_t does not eliminate it. The next most important part of the secondary capacity is the component C_w, which is built up from the capacity between the successive layers of secondary winding connected effectively in series. The dielectric of this condenser consists of the insulation between layers, while the copper wires of the successive layers form the plates.

In addition to these two main components there is a certain amount of capacity C_d represented by condensers, one plate of which is the end turns of the successive layers of the secondary, while the other plate is the end of the armature core. This capacity is "distributed" in the strict sense of the word, and the effect of its component parts upon the entire machine performance must be weighted in proper proportion, depending upon the voltage which is applied to each successive layer.

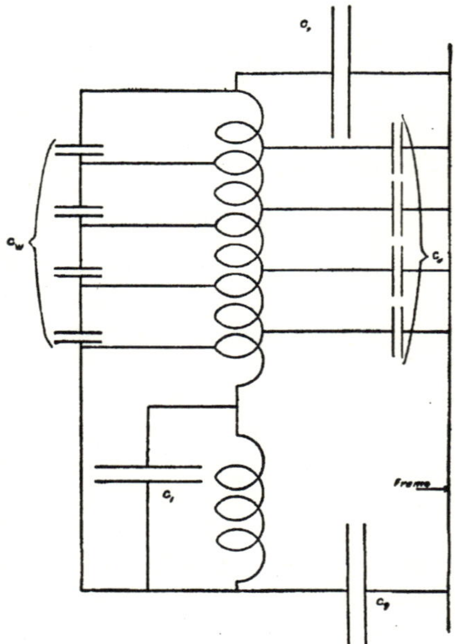

FIG. 23.—*Schematic diagram of capacity distribution in magneto.*

The equivalent capacity C_2 which is desired for use as the contribution from the secondary winding in any of the models discussed above is that which would contain at the terminal voltage of the winding the same amount of electrostatic energy as is actually stored in the various portions of the dielectric, and the use of such an equivalent condenser is accurate except to the extent that there may be local oscillations of higher frequency, involving only portions of the winding and of the corresponding condensers.

1. BY IMPULSIVE IMPEDANCE

It is possible to obtain the equivalent total secondary capacity from measurements on the magneto by observing the crest voltage obtained by interrupting a known current through the primary with the primary condenser removed. This measurement should be made with a number of different values of capacity added to the secondary terminals, and the corresponding values of the impulsive impedance Z should be computed. If $\frac{1}{Z^2}$ is then plotted as ordinate against the value of the added capacity as abscissa, the results will be found to fall approximately upon a straight line, the intercept of which with the axis of abscissae will give the equivalent secondary capacity previous to the addition of the condenser.

2. BY COMPUTATION

The capacity through the winding C_w may be computed approximately from the dimensions of the winding if the dielectric constant of the insulating material is known. In this computation it is sufficiently accurate to assume the condenser to be built up of infinite parallel plates whose thickness is that of the dielectric between nearest points of the wires in adjacent layers and whose area is the product of the circumference of the mean turn multiplied by the width of the winding. Dividing this by the number of layers to allow for series connection gives the effective value which the capacity would have if the voltage gradient throughout the width of each condenser were uniform. In the actual winding, however, the dielectric at one end of the layer is subjected to a voltage $2e$, while that at the other end is 0, if e is the voltage per layer, and the total energy stored in the condenser is greater than $\frac{1}{2} C e^2$ because of the nonuniform distribution of voltage along each layer of dielectric. Multiplying the results obtained as indicated above by the factor $\frac{4}{3}$ corrects for this condition and gives the value which should be used in conjunction with the capacity C_t obtained by other methods to give the total capacity of the winding.

3. BY RESONANCE

Another very convenient method for measuring the capacity of magneto windings is that indicated in Fig. 24, where the coil A is supplied from a suitable source with currents of radio frequency (100 000 to 500 000 cycles per second). The oscillating circuit

B, loosely coupled to *A*, is tuned to resonance, as indicated by a suitable hot wire instrument (*A*). The terminals of the winding *w* whose capacity is to be measured are then connected in parallel with the variable condenser C_0, and this latter is then readjusted to again establish resonance. The decrease in the condenser C_0 is, of course, equal to the value of the capacity which was connected in parallel with it by the insertion of the winding. For frequencies of the magnitude indicated, the inductance of the magneto winding offers a practically infinite reactance, and only a negligible amount of current flows spirally through the winding around the core. The main part of the radio frequency current flows through the dielectric from layer to layer of condenser C_w and also supplies C_t. In using this method it is well to take readings at several different frequencies, so that if one of them coincides with a natural overtone of the coil the resulting error will be apparent from the lack of agreement with the values obtained at the other frequencies.

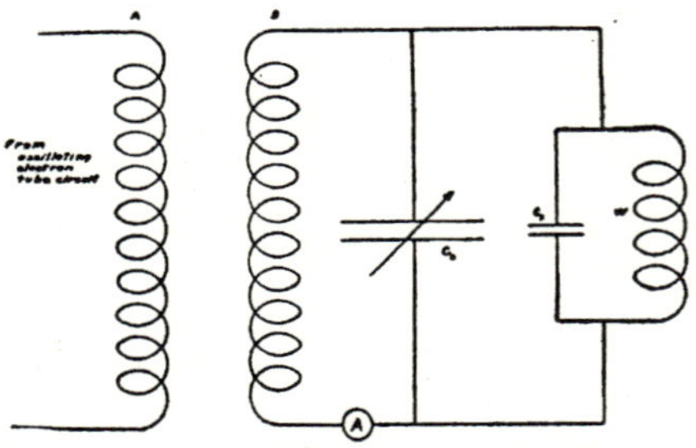

FIG. 24.—*Circuits used in measuring capacity, C_x of magneto winding W at radio frequencies*

If the grounded terminal of the primary winding can be separated from the core, the entire magneto may be regarded as having three terminals: (1) The high-tension terminal of the secondary; (2) the insulated ground terminal of the primary; (3) the frame. Any two of these terminals may be used as the terminals of a condenser to connect in parallel with C_0, while the third terminal may either be left insulated or may be connected to one of the other two. This system thus gives six possible combinations, any three of which are sufficient to determine the three capacities C_w, C_t, and C_g (C_g being the capacity between the inner layer of the primary winding and the frame). The check values obtained from the other three combinations will usually be found to agree to a few per cent.

It should be borne in mind that during these radio frequency measurements the total voltage is distributed between the layers by their respective capacities and that the voltage from any one layer to the next is uniform throughout the width of the layer, and consequently the radio frequency values must be increased by the factor $\frac{4}{3}$ for the same reason as the computed values.

4. BY A. C. BRIDGE

Still another method for measuring C_t is obtained by the alternating current bridge method described in detail below. (See Fig. 25.) With this bridge the charging currents flowing from the winding to the frame are led off through the condenser C_6 and in effect measured by it. This method gives the quantity C_t combined with an effective value of C_d weighted in proportion to the voltage applied to each element of C_d, but does not give a measure of C_w. Probably the most satisfactory values can be obtained by combining C_t obtained by this method with C_w obtained at radio frequency.

The following table gives some typical data on secondary capacity obtained by the various methods outlined in the case of two shuttle type magnetos:

Capacity	Observed at radio frequency	Computed from dimensions	A. C. bridge at 1000 cycles
Magneto D-82:			
C_t+C_d	45.3	...	46.2
C_w	$\frac{4}{3}\times 23.3$	$\frac{4}{3}\times 27.1$	...
C_g	253.	...	...
Magneto D-81:			
C_t+C_d	43.4	...	40.7
C_w	$\frac{4}{3}\times 28.5$	$\frac{4}{3}\times 27.1$	...
C_g	213.	...	...

In the case of D-82 the total effective capacity, referred to the secondary, obtained by summing the various components (including the primary condenser and secondary leads) was found to be 197 $\mu\mu f$, while the value obtained by extrapolating the line obtained by plotting $\frac{1}{Z^2_m}$ against added capacity gave 204 $\mu\mu f$.

Resistance and Inductance Measurements

The double-coil and closed-coil models involve too many constants to lend themselves readily to the determination of the numerous inductances and resistances needed by the second method outlined above. The curve of $\frac{1}{Z^2}$ vs. capacity mentioned above should have, according to the single-coil model (see equation 27), a slope equal to $\frac{e^{-2\frac{\alpha}{\beta}\tan^{-1}\frac{\beta}{\alpha}}}{L}$, and this combination of the resistance and inductance may be used in computations based upon the single-coil model or may, in fact, be used as the reciprocal of an equivalent inductance in a single-coil model of zero resistance to represent with fair accuracy the phenomenon which takes place. It might be thought that measurements of the maximum negative crest of the voltage wave, in addition to those of the positive crest, would give a value for the damping of the wave, and hence would serve to measure the resistance of the equivalent single-coil model. In the magnetos tested, however, the distortion of the base of oscillations by the eddy currents, as indicated by the term involving $e^{m_1 t}$ of equation (67) is so great that the values of resistance obtained by the method outlined are very much greater than those observed by the alternating current bridge described below and give a rather misleading indication of the amount of energy dissipation.

1. BY BALLISTIC GALVANOMETER

Measurements with reversed direct current using a ballistic galvanometer to measure the changes in flux involved serve to indicate the primary and secondary inductance and mutual inductance for such slow changes in flux, and these values are suitable for insertion in the closed-coil model, since the tertiary circuit takes care of the decrease in these inductances and the increase in resistance which occur at higher frequency. The values obtained ballistically, however, are not satisfactory for use directly in the double-coil or single-coil models. A null ballistic method may be used to obtain the coefficient of coupling k between the primary and secondary windings for use in any double coil model.[19] While the values of k obtained by this method are somewhat in error as the result of the nonuniform current distribution in the secondary winding under operating conditions, the capacity measurements indicated above show that the component C_d of capacity which causes this nonuniform current is relatively small, and the error introduced from this cause is probably negligible compared with the other errors which

[19] A. P. P. Report No. 16, National Advisory Com. for Aeronautics, Report No. 58 II.

are inherent in setting up a double-coil model equivalent to any given magneto. The changes in inductance resulting from changes in frequency are mainly dependent upon changes in the flux distribution within the iron core itself, and it is probable that they have relatively little effect upon the coefficient of coupling between the primary and secondary winding, although both L_1, and L_2 and M are individually greatly affected by an increase in frequency.

2. BY A. C. BRIDGE

The methods of measuring resistance and inductance by means of some modified form of Wheatstone bridge supplied by alternating current are rapid and convenient and may be applied to the determination of the constants for the various types of magneto models discussed above. Any such measurement, in fact, indicates the values of resistance and inductance of an equivalent circuit which, when supplied with sustained alternating current of the frequency used, shows the same magnitude and phase relation of current as does the actual apparatus tested. During the operation of a magneto in the normal manner its winding and core are subjected to an impulse which bears a close resemblance to a single half wave of sustained alternating current, and it consequently seems a reasonable assumption that the value of resistance and inductance applicable in computing such an impulse should be at least approximately the same as those observed for an alternating current of corresponding frequency.

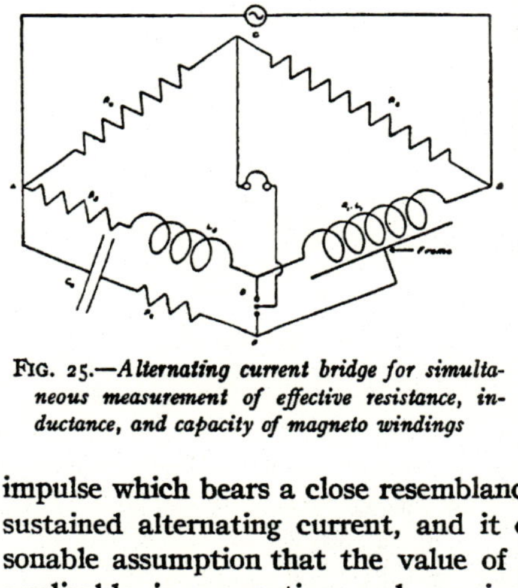

FIG. 25.—*Alternating current bridge for simultaneous measurement of effective resistance, inductance, and capacity of magneto windings*

The type of alternating current bridge which has been found convenient for such measurements is shown in Fig. 25, in which one winding, preferably the high-voltage coil, is connected as one of the four arms of a Wheatstone bridge, the primary condenser being disconnected entirely. In addition to the four normal arms, an additional circuit is connected from point A through variable condenser C_6 and a variable resistance R_6 to the frame of the magneto, which has been disconnected from the end of the primary winding with which it is normally in contact. Inductance L_5 and resistance R_5 are then adjusted until a telephone or other suitable detector connected between points C and D shows a balance. The detector is then connected between points C and F and C_6 is adjusted for balance. When by several repetitions of this process a simultaneous balance has been obtained on both of these points, we have to a close degree of approximation the relation

$$R_1 + j\omega L_1 = \frac{R_3}{R_4}(R_5 + j\omega L_5) \tag{82}$$

from which R_1 and L_1 for each frequency can be obtained.

Equation (82) is rigorous if C_d is zero, so that the entire capacity between the winding and frame can be considered as concentrated at the ends as C_t and C_g, respectively. If this is not the case, but if the condensers C_d and C_t are free from energy loss, they may be combined to form an equivalent condenser C_k, located at a fraction K of the total length of the winding from the terminal B. Equation (82) then holds if the right-hand member is multiplied by the correction factor

$$\frac{1}{1 + K\frac{Z_5}{Z_6}} \tag{83}$$

where Z_5 and Z_6 are the impedances of arms 5 and 6, respectively, expressed in vector notation. K is given by the equation

$$K = \frac{R_3 R_6}{R_4 R_1'} \tag{84}$$

Where R_1' is that part of the coil resistance due to the copper winding only. Unfortunately, however, the condenser C_t usually has an appreciable energy loss (corresponding to a power factor of about 0.05), and a certain amount of resistance must be inserted in arm 6 to balance this loss. This resistance is often so large that the value of K computed from (84) is very seriously in error. The amount and location of the capacity C_d can be estimated roughly from the arrangement and dimensions of the windings and in the coils thus far studied is found to be negligible in its effect on equation (82).

The magnitude of the condenser C_k, which is equivalent in the bridge circuits to the combined effect of C_d and C_t, is given by

$$C_k = \frac{R_4}{R_3} \cdot \frac{C_6}{1 - K} \tag{85}$$

For use in the various artificial models, however, we desire the value of condenser, which, if connected from the secondary terminal to ground, would receive the same energy as is actually stored in the distributed capacity. This value is given by

$$C_k' = (1-K)^2 C_k = \frac{R_4}{R_3} C_6 (1-K) \tag{86}$$

In case C_d can be neglected, K becomes zero and (85) and (86) reduce to

$$C_t = \frac{R_4}{R_3} C_6$$

It will be noticed that the capacity C_g, discussed above, which may exist between the inner layer of the primary winding and the core, does not affect the measurement in any way, since it is not subjected to any difference of potential when the balance has been obtained. The component of capacity denoted above by C_w from layer to layer of the winding does not contribute to the flow of current through the arm C_6, but flows through arm 5 of the bridge. The results of equation (82) must, therefore, be corrected for this effect also by the following equations:

$$R = \frac{R'}{(1+\omega^2 L'C)^2 + \omega^2 C^2 R'^2} \tag{87}$$

$$L = \frac{L'(1+\omega^2 L'C) + CR'^2}{(1+\omega^2 L'C)^2 + \omega^2 C^2 R'^2} \tag{88}$$

In these equations R and L are the true resistance and inductance of the winding as they would be measured if the capacity C_w were absent, while R' and L' are the actual observed resistance and inductance as computed from equations (82) and (83) above. These equations are the converse of the well-known equations:

$$R' = \frac{R}{(1-\omega^2 LC)^2 + \omega^2 C^2 R^2} \tag{89}$$

$$L' = \frac{L(1-\omega^2 LC) - CR^2}{(1-\omega^2 LC)^2 + \omega^2 C^2 R^2} \tag{90}$$

which express the effective resistance and inductance of a coil which is shunted by a condenser.

Figs. 26, 27, and 28 show typical results obtained on three shuttle-type magnetos by this method. The dotted curves give

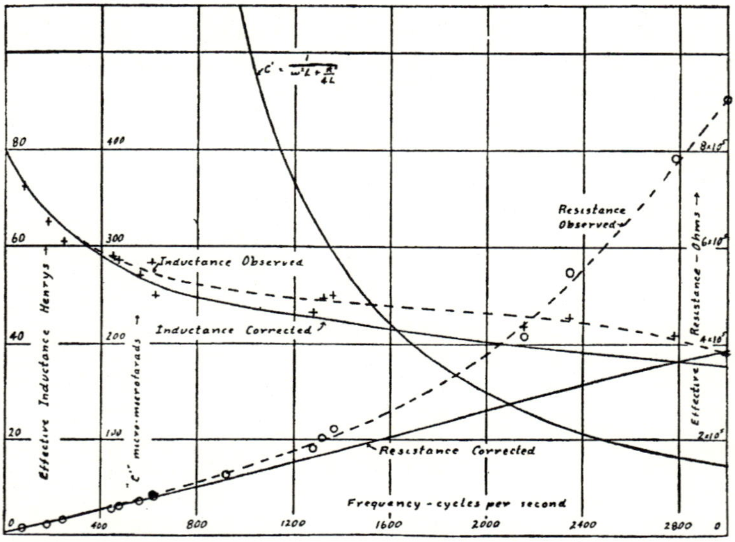

FIG. 26.—*Effective resistance and inductance of D-82 (laminated pole) magneto as measured on alternating current at various frequencies*

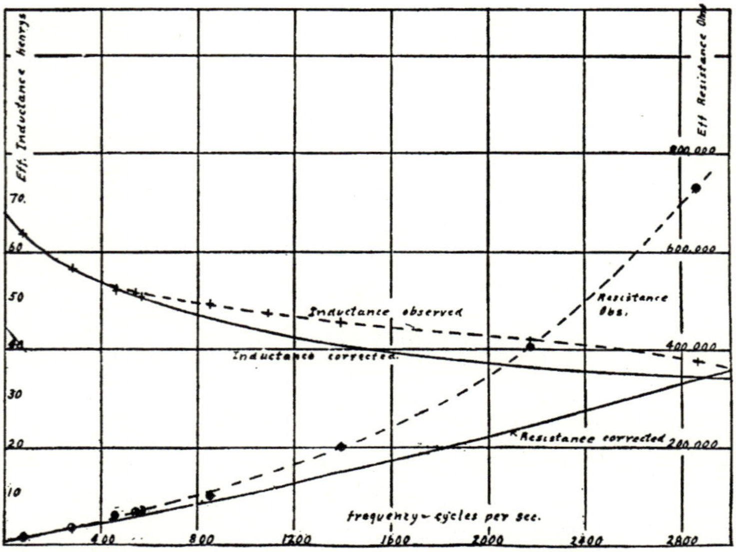

FIG. 27.—*Effective resistance and inductance of magneto having a D-81 magneto armature in a D-82 (laminated) pole structure*

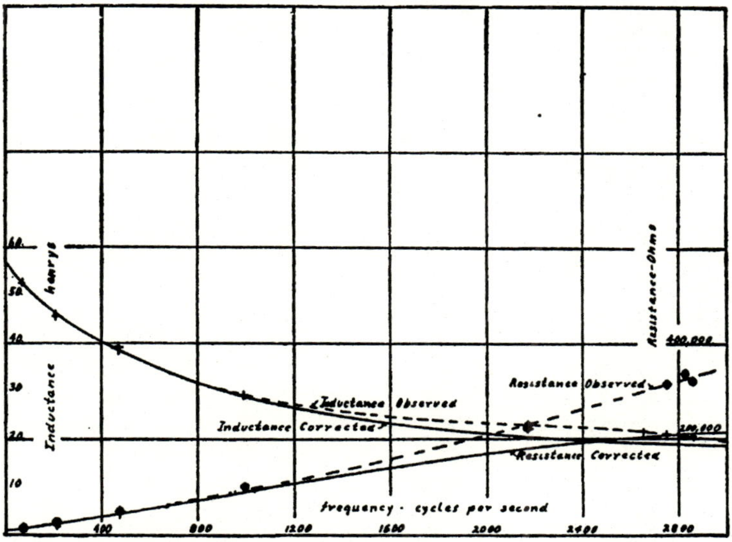

FIG. 28.—*Effective resistance and inductance of D–81 magneto (solid pole pieces)*

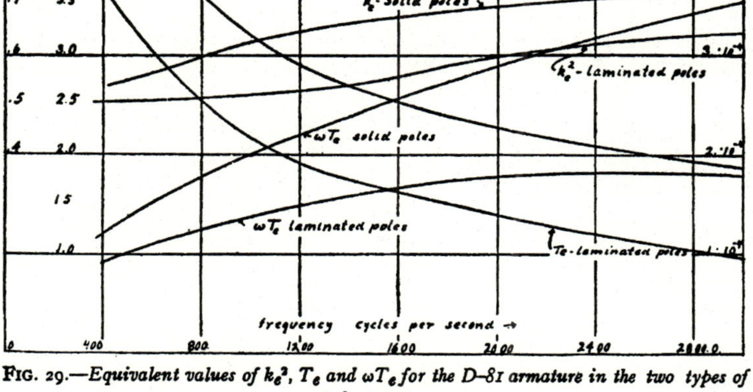

FIG. 29.—*Equivalent values of k_e^2, T_e and ωT_e for the D–81 armature in the two types of pole structure*

the observed resistance and inductance for various frequencies, while the solid curves show the corresponding values after correction has been made for the capacity C_w, which in these cases amounted to 31, 38, and 38$\mu\mu f$, respectively. It will be noted that the inductance first decreases quite rapidly from the values observed on direct current and then maintains a fairly constant value as the frequency is further increased. The effective resistance increases very rapidly to a value which is many times that of the copper winding and indicates forcibly the importance of the iron losses in effecting the dissipation of energy in such an apparatus.

To apply the results of measurements by this method to the single-coil model, it is necessary to select the values of effective resistance and inductance which correspond to the particular frequency, a half wave of which most nearly coincides with the impulse corresponding to the operation of the model. This can be done either by a series of successive approximations or more readily by plotting (see Fig. 26) the auxiliary quantity

$$C' = \frac{1}{\omega^2 \left(L + \frac{R^2}{4L} \right)} \qquad (91)$$

C' may be defined as the capacity which, substituted in equation 23, will give as the resulting frequency $\frac{\beta}{2\pi}$ of a single-coil model the value $\frac{\omega}{2\pi}$. Hence, when the total capacity connected to the magneto is known in any specific case, the frequency corresponding to this value of C' on the auxiliary curve will be that for which R and L should be taken.

If the primary condenser is not disconnected, it produces a very marked effect upon the effective resistance and inductance, as would be expected from equations (89) and (90). Fig. 30 shows the corresponding variation of resistance and inductance with frequency and indicates that a resonance point is attained at about 2300 cycles.[20]

[20] In cases where the damping is appreciable three different frequencies of oscillation are of importance, as follows: A circuit of the single-coil type oscillating freely has a frequency, $f = \frac{1}{2\pi}\sqrt{\frac{1}{LC} - \frac{R^2}{4L^2}}$; if alternating current is applied to the terminals of the condenser of such a system, the effective inductance has the value 0 at the slightly different frequency, $f = \frac{1}{2\pi}\sqrt{\frac{1}{LC} - \frac{R^2}{L^2}}$; and the effective resistance has a maximum at a still different frequency, $f = \frac{1}{2\pi}\sqrt{\frac{1}{LC} - \frac{R^2}{2L^2}}$.

The results of A. C. bridge measurements may be correlated with the closed-coil magneto model by computing the values of k_e and T_e which a tertiary circuit must have in order to change the resistance and inductance from their D. C. values to the values which are observed at any particular frequency of A. C. The effect of such a tertiary circuit upon the main coil is indicated by equations

$$R_1' = R_1 + \frac{\omega^2 k_e^2 T_e L_1}{1 + \omega^2 T_e^2} \tag{92}$$

$$L_1' = L_1 \frac{(1 + \omega^2 T_e^2(1 - k_e^2))}{1 + \omega^2 T_e^2} \tag{93}$$

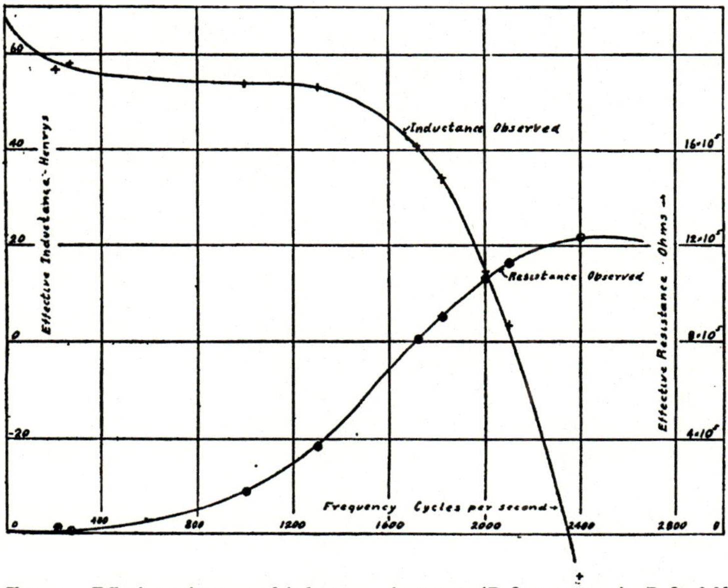

FIG. 30.—*Effective resistance and inductance of magneto (D–81 armature in D–82 field with primary condenser connected*

which result immediately from (79), and, conversely, it is possible to compute for each frequency the values of k_e and T_e by the following formulas which are derived from equations (92) and (93):

$$T_e = \frac{L_1 - L_1'}{R_1' - R_1} \tag{94}$$

$$k_e^2 = \frac{L_1 - L_1'}{L_1} + \frac{(R_1' - R_1)^2}{(L_1 - L_1')L_1\omega^2} \tag{95}$$

The resulting values of R_1, L_1, and k_e, T_e for any frequency constitute the constants of the corresponding closed-coil model. (Fig. 29.)

VI. EXPERIMENTAL CONFIRMATION OF MODELS

Probably the most searching qualitative tests of the various models discussed above is offered by the measurements at the National Physical Laboratory of the actual wave form during period 2. A comparison of Fig. 9, taken from their results, with Fig. 31 shows that the general character of a damped sinusoidal impulse can be accounted for by the single-coil model (Curve II), and that the shift of the base line of this impulse to a positive exponential curve and the consequently great reduction in the negative maximum is explained by the closed-coil model (Curve III), and that the high-frequency ripple with its relatively small amplitude can be explained by the double-coil model (Curve I).

The quantitative comparison of the models with an actual magneto is offered by tests made at the Bureau of Standards of two shuttle type Berling magnetos, models D–81 and D–82, respectively. These types are identical except that model D–82 has laminated pole pieces while D–81 has solid iron castings, and in certain of the measurements (e. g., Fig. 27) the D–81 armature was used in both frames to eliminate all other possible differences in winding. Measurements of the capacity of these magneto windings made by the radio frequency measurements are given above (p. 455), and the results of the A. C. bridge measurements made on the secondary winding of resistance, inductance, and capacity to ground are those shown in Figs. 26, 27, and 28. Throughout the alternating-current measurements the voltage applied to the bridge was varied in proportion to the frequency, so that the magnetic circuit operated at approximately the same flux density at all frequencies. Although 500 volts were applied at 3000 cycles, the flux was necessarily much less than that occurring in normal operation and corresponded to that produced by a primary current at break of 0.03 ampere. This latter value was used in the crest-voltage measurements.

The results of the experimental work are best expressed in terms of the "impulsive impedance" or ratio of the crest voltage to the current at break. While this impedance, like any other constant of the system, may be referred to either the primary or secondary circuits, it is for most purposes more convenient to use the hybrid value given by the quotient of secondary crest voltage, divided by primary current at break. This quantity Z_m is related

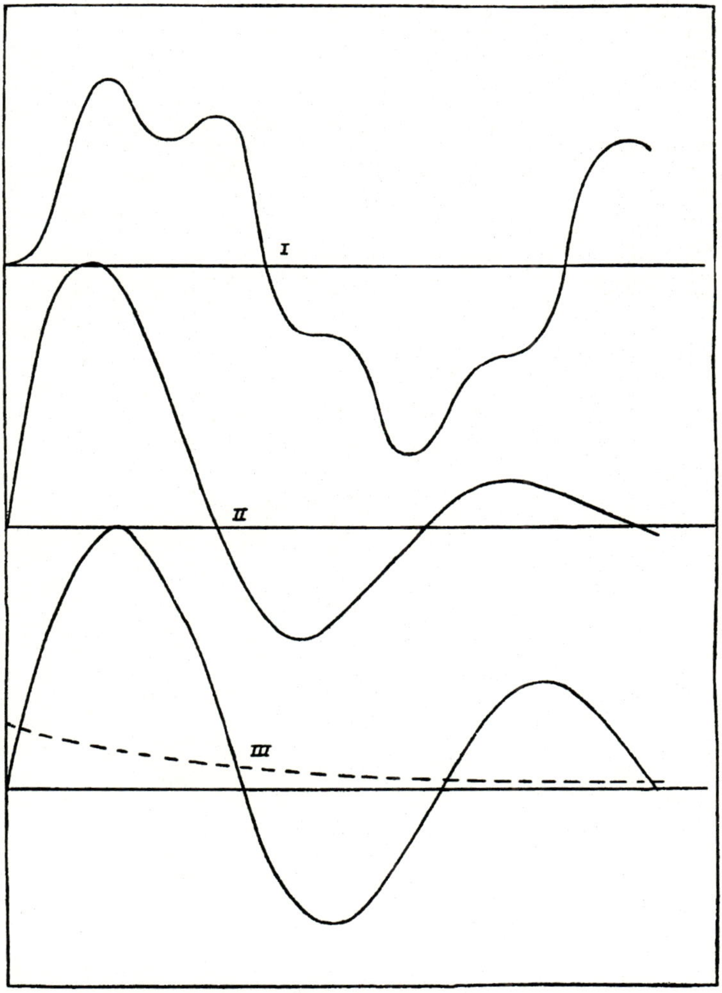

FIG. 31.—*Voltage wave forms of various models: I, double-coil model; II, single-coil model; III, closed-coil model*

to the primary (Z_1) and secondary (Z_2) impulsive impedance by the equation

$$Z_2 = n\ Z_m = n^2 Z_1 \tag{96}$$

where n is the ratio of turns of the magneto.

Fig. 32 shows the variation of this quantity when different condensers were connected to either the primary or secondary of the magneto. It will be noted that the points indicated as circles which correspond to various additions of capacity to the primary windings lie on the same curve as those indicated by crosses which correspond to the addition of secondary capacity, this agreement constituting a very satisfactory justification of the utility of referring electrical quantities from one side to the other of the transformer, as was done in developing the single-coil model.

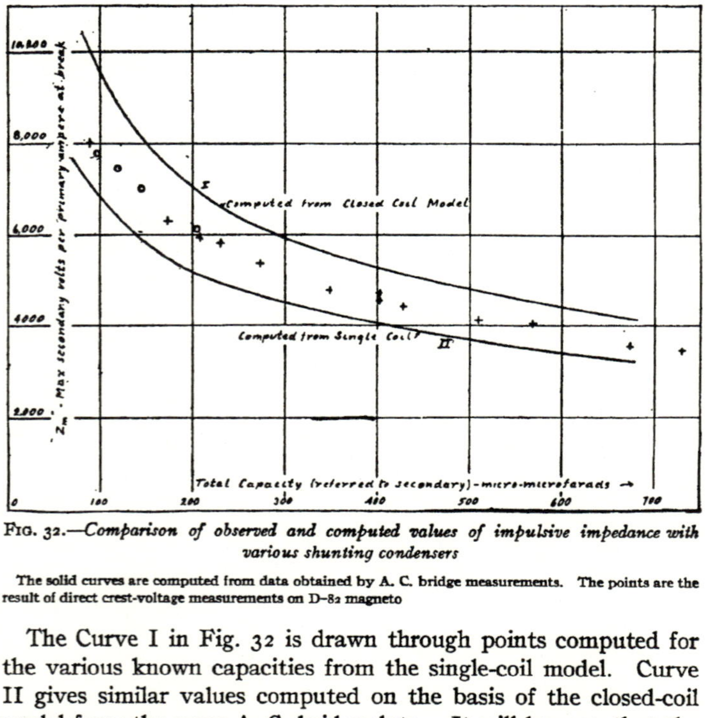

FIG. 32.—*Comparison of observed and computed values of impulsive impedance with various shunting condensers*

The solid curves are computed from data obtained by A. C. bridge measurements. The points are the result of direct crest-voltage measurements on D-82 magneto

The Curve I in Fig. 32 is drawn through points computed for the various known capacities from the single-coil model. Curve II gives similar values computed on the basis of the closed-coil model from the same A. C. bridge data. It will be seen that the discrepancy in both cases, while greater than the error of measurement and fairly constant, is by no means large when one considers that the points and the curves were obtained on entirely distinct experimental bases, there being no data common to the two sets of observations.

Fig. 33 gives a similar comparison when various shunting resistances were connected in parallel with the secondary of the magneto. The crosses represent the values of Z_m directly observed. The solid curve is computed from the single-coil model using the values of L and R, observed on the A. C. bridge at a frequency of 2200 cycles, which is the natural frequency of oscillation of the model when unshunted. When the shunting is very heavy, the observed points depart very considerably from the curve thus computed as might be expected from the slowing down of the oscillations by the shunt. If in the computation for any point the values of L and R corresponding to the actual frequency of the oscillation are used in the single-coil model, the values of Z_m indicated by the

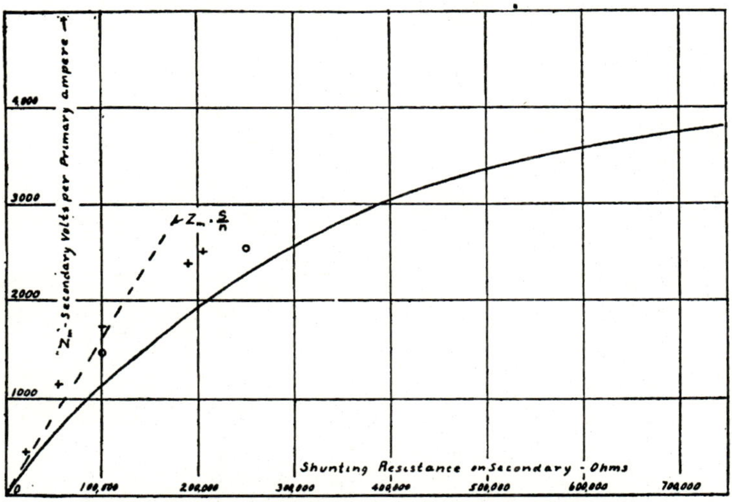

FIG. 33.—*Comparison of observed and computed values of impulsive impedance with various shunting resistances*

The solid curve is computed from the constants of a single-coil model fitting the A. C. bridge data at 2200 cycles. The circles are computed for a single coil model having the A. C. constants of the actual magneto at the particular frequency of oscillation when shunted. Crosses indicate the result of direct crest-voltage measurements on D-82 magneto

circles are obtained, which are seen to agree much more closely with the observed values than does the solid curve. In cases where the oscillations are overdamped the values of L and R corresponding to zero frequency may be used.

Fig. 34 shows the variation of the secondary crest voltage with the magnitude of the primary current at break. The straight-line relation indicates that in spite of the presence of iron in the magnetic circuit the variation of Z_m with current is relatively slight. This constancy of the inductance and hence of Z_m is of value inasmuch as it is not practicable to make the A. C. meas-

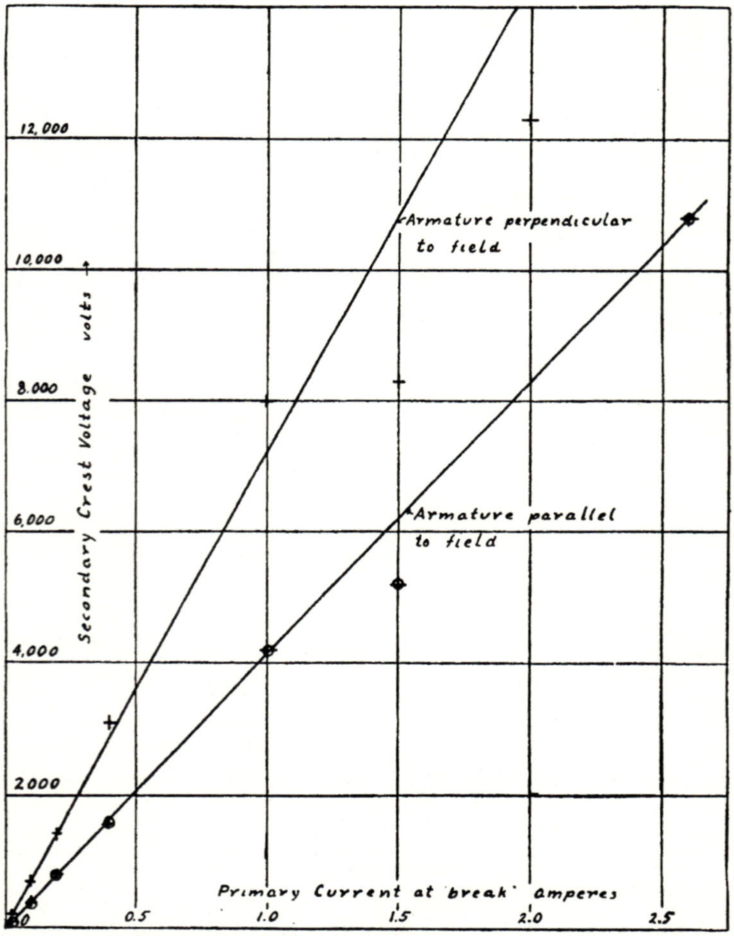

FIG. 34.—*Data showing approximate proportionality between crest voltage and primary current. (The low values at 1.5 amperes are the result of sparking at the breaker points)*

urements with as large current values as occur in actual operation of the magneto. If this were attempted, the resultant heating of the winding by the great energy loss which would occur would be excessive.

VII. CONCLUSIONS

The experimental results just described indicate that the various models proposed in this paper do not depart widely in their performance from the actual magneto, and it is hoped that the single-coil model in particular, because of its simplicity, may be found useful in designing apparatus of this type. There is a great need for experimentation by the methods suggested on a wider variety of magnetos of different types in order to confirm the validity of the models for other types of construction than the conventional shuttlewound armature and to indicate the order of magnitude of the quantities involved in typical cases. Measurements already made show that the quantity T_e is a definite quantitative measure of the effect of eddy currents, and a study of the values of this quantity in magnetos having different thickness of laminations would be of great value.

WASHINGTON, December 1, 1920.

Spotted something that isn't quite right?

Let us know and we'll fix it for future reprints. And as a thank you for your diligence, we'll send you a free Sothis book of your choice.

publisher@sothis.press

This page intentionally left blank